CHEMICAL ENGINEERING
A Review for the P.E. Exam

To Bunny

PREFACE

The West Virginia College of Graduate Studies offers the Master's degree in Chemical Engineering for the employees of the many chemical companies near Charleston, West Virginia. As a public service, the College offers a review course for chemical engineers who wish to become registered. I have taught this course eleven times.

Like many college professors, I could not find a suitable text. So I taught the course from notes that I changed (improved?) each time that I taught. With yearly after-the-test comments from most course participants, the notes have reached the present form. I am satisfied with them and the students are satisfied with them, so it is time to have them published.

I am indebted to Mrs. Reva Dunn for her many months of typing, even though she now hates equations; to Mrs. Margery Edington for many hours over a hot photocopy machine; to Mr. Robert Oneacre for his weeks of hunching over a drawing table doing the figures; to Ms. Jeannette Stiefel and Ms. Margaret Comaskey for making me look good with their editing; and to the West Virginia College of Graduate Studies for the timely financial grant. I thank the students, for it is they who make it a joy rather than a job. Above all, I am grateful to Bunny, who makes it all worthwhile.

William E. Crockett
P.E. #6853 (WV)

Institute, West Virginia
August 1985

vii

CONTENTS

CHEMICAL ENGINEERING
A Review for the P.E. Exam

INTRODUCTION

This book will help you prepare for the Chemical Engineering Principles and Practice of Engineering Examination (P. E. Exam). Although it will increase your chances of passing the exam, I hope you will also relearn neglected parts of your undergraduate chemical engineering. I have included only those subjects and concepts likely to occur on the test.

To make the book useful during the exam, I have:

- used the table of contents for each chapter as a topic index. I assume that upon reading a problem, you will be able to identify the type of problem. Scanning the table of contents of the appropriate subject chapter should lead you to the proper section of the chapter.
- defined the nomenclature in the text rather than in a separate section, which should be more convenient during the test.
- followed the same topic sequence as used in McCabe and Smith's *Unit Operations of Chemical Engineering* for the chapters on mass transfer, heat transfer, and fluids. The thermodynamics topic sequence follows that of Smith and Van Ness' *Introduction to Chemical Engineering Thermodynamics*. Both of these texts are standards; many of you used them when you were undergraduates. The process design, kinetics, and economics chapters follow schemes that I have found useful.

THE TEST

The P. E. Exam is an 8-hr test, administered in two 4-hr sessions. In each session you are to work any 4 of 10 problems. The approximate number of problems in each subject is

Thermodynamics	3	Chemical kinetics	4
Process design	4	Fluid flow	2
Mass transfer	4	Engineering economics	1
Heat transfer	2		

A test booklet contains all the problems. You must put all the solution steps and calculations on the graph-paper solution pages provided in the booklet.

You should take pens or pencils with you to the test. You may use a hand-held calculator, but it must be battery operated. Take extra batteries to be safe. Since the test is open book, reference texts, handbooks, and math tables are permissible. But loose pages are not permitted unless they are kept in a binder or notebook.

People grade the test. Partially correct solutions receive partial credit. Do not give the graders any opportunities to misunderstand your solutions. If any part of a problem is stated ambiguously, state (and underline) your interpretation of the meaning.

REFERENCE TEXTS

I shall suggest reference books that I prefer, but it would be better if you choose books with which you are familiar. My reference texts are

W. L. McCabe and J. C. Smith, *Unit Operations of Chemical Engineering*, 3rd ed., McGraw-Hill, New York, 1976. An excellent reference with a great index.

J. M. Smith and H. C. Van Ness, *Introduction to Chemical Engineering Thermodynamics*, 3rd ed. McGraw-Hill, New York, 1975. Comprehensive and easy to understand.

F. Kreith, *Principles of Heat Transfer*, 3rd ed., Harper & Row, New York, 1973. Comprehensive treatment of all types of heat transfer.

O. Levenspiel, *Chemical Reaction Engineering*, 2nd ed. Wiley, New York, 1972. Widely used with excellent example problems.

E. L. Grant, W. G. Ireson, and R. S. Leavenworth, *Principles of Engineering Economy*, 7th ed. Wiley, New York, 1982. The standard text.

R. H. Perry, *Chemical Engineers' Handbook*, McGraw-Hill, New York, 1973, latest edition. Excellent source for data but written by a committee.

J. H. Keenan and F. G. Keyes, *Thermodynamic Properties of Steam*, Wiley, New York, 1954.

CRC Standard Mathematical Tables, 20th ed., The Chemical Rubber Company, Cleveland, Ohio, 1972.

I know several engineers who took only Perry and this book to the test. They did fine, but several reference books may provide a psychological benefit.

PREPARING FOR THE TEST

The test is easier and less frightening if you have prepared properly. Here are some suggestions:

- Start preparing at least several months before the test. Tests are usually given in mid-April and at the end of October of each year.

- Decide upon and get a copy of the reference books that you prefer.
- Read a chapter in this book, stopping before the problems. Each chapter deals with one subject; they do not have to be studied in sequence. You may wish to do your most difficult subject first.
- Page through the reference book you have chosen for the subject and become familiar with the contents.
- Returning to this book, read all the problems provided at the end of the chapter, and find the easiest. You should do this on the test too. The secret to success is finding and solving the easiest problem first. It builds confidence and saves test time for more difficult problems.
- Determine what problem-solving strategy you will use. This is usually done mentally as you read and think about each problem. Place the problem in a broad category and then narrow the category again and again; for instance, heat transfer, condensing vapor to heat a liquid, double-pipe exchanger, log mean temperature difference, and dirt factor. By using this strategy you start to "see" the needed equations or, at least, to know which areas to look for in the table of contents of the subject chapter.
- Compare your strategy with the strategy given in the chapter. If you are not used to solving problems using this step, you may find it beneficial.
- Solve the chosen problem. Treat it just as if it were an actual test problem. While both preparing for and taking the test:
    Time yourself.
    Sketch the process, if appropriate, to make the problem clearer.
    Write down the basis for you calculation; you will be less likely to change bases inadvertently.
    Write the units with each number. You will prevent some simple mistakes this way.
    List the reference for any data or equations used, so the grader can check your source.
- Compare your solution with the one given at the end of the chapter.
- Continue with the other problems, easiest to hardest.
- Continue with the other chapters, possibly hardest to easiest. If you should run out of preparation time, at least you have studied those subjects on which you most need refreshing.

Remember that you do not have to be proficient in all subjects to be able to work 8 problems out of 20. Eight correct solutions would give you a perfect score, and you do not need a perfect score to pass.

You may wish to finish your test preparation with a sample examination marketed by the P.E. Exam makers—the NCEE, National Council of Engineering Examiners.† The cost is about $20.

---

† P. O. Box 1686, Clemson, South Carolina 29633–1686.

THERMODYNAMICS

J. M. Smith, and H. C. Van Ness, *Introduction to Chemical Engineering Thermodynamics*, 3rd ed., McGraw-Hill, New York, 1975.

I. DEFINITIONS

A. Fundamental Quantities

1. Time

The fundamental unit of time is the second.

2. Length

The fundamental unit of length is the meter. A foot is 0.304 meter (m).

3. Mass

The fundamental unit of mass is the kilogram. The pound mass (lb_m) is 0.454 kilogram (kg). Mass is the measure of quantity of matter, while weight refers to the force exerted by gravity on mass. An object's weight may vary from place to place but its mass remains constant.

4. Force

The fundamental unit of force is the newton (N), which is the force which, when applied to a one kilogram mass, will cause an acceleration of one meter per second. This unit comes from Newton's second law of motion

$$F = ma/g_c$$

where g_c is a proportionality constant. Note that this definition provides a relationship among four fundamental quantities—time, length, mass, and force. Thus the units of these quantities determine the value of g_c. The newton is defined such that

$$g_c = 1 \ \frac{(kg)(m)}{(N)(sec^2)}$$

For the English system, using pounds force (lb_f) and pounds mass (lb_m),

$$g_c = 32.174 \ \frac{lb_m \ ft}{lb_f \ sec^2}$$

5. Temperature

The fundamental unit of temperature is the degree Celsius (°C). It is related to the absolute temperature K on the Kelvin scale, by

$$t(°C) = T(K) - 273.15$$

The English unit is the degree Fahrenheit (°F), which is related to the absolute temperature °R on the Rankine scale by

$$t(°F) = T(°R) - 459.67$$

The relationship between the two temperature scales is

$$t(°F) = 1.8t(°C) + 32$$

B. Secondary Quantities

1. Volume

The amount of material being considered affects the volume. But specific volume is independent of the total amount of material since it is defined as volume per unit mass or volume per unit mole. The density ρ is just the reciprocal of specific volume.

2. Pressure

The pressure P on a surface is the force exerted normal to the surface per unit area of surface. The pascal, newtons per square meter, is used in the SI system while psi or pounds force per square inch is used in the English system. Pressures are usually measured in terms of a height of a column of fluid under the influence of gravity. Since

$$F = mg/g_c,$$

if the mass of the column of fluid is m,

$$m = Ah\rho$$

then

$$F/A = P = h\rho g/g_c$$

Remember that many of the methods of measuring pressure give the difference between the measured pressure and the surrounding atmospheric pressure, but all thermodynamic calculations require the use of absolute pressure.

3. Work

When a force acts through a distance, we have work being done. The quantity of work is defined by

$$dW = F\,dl$$

where F is the force component acting in the displacement dl direction. With this definition, work accompanies a change in fluid volume. For the expansion or compression of a fluid in a cylinder caused by a piston movement, the force exerted by the piston on the fluid is given by the product of the piston area and

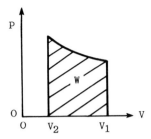

FIGURE T1. *PV* work.

the fluid pressure. The piston displacement is given by the fluid volume divided by the piston area. So,

$$dW = PA\,d\frac{V}{A} = P\,dV$$

for constant A. Then

$$W = \int_{V_1}^{V_2} P\,dV$$

as shown in Figure T1. In the SI system, the unit of work is the newton meter, which is called a joule (J). In the English system, the foot-pound force (ft lb$_f$) is often used.

4. Energy

For a mass m acted upon by a force F, the relationship was given by $F = ma/g_c$, so

$$dW = \frac{ma}{g_c}\,dl$$

Now acceleration is $a = du/dt$, where u is the body's velocity. Thus

$$dW = \frac{m}{g_c}\frac{du}{dt}\,dl = \frac{m}{g_c}\frac{dl}{dt}\,du$$

But velocity is $u = dl/dt$, so

$$dW = \frac{m}{g_c}u\,du$$

Integrating for a change in velocity from u_1 to u_2,

$$W = \frac{m}{g_c}\int_{u_1}^{u_2} u\,du = \frac{m}{g_c}\left(\frac{u_2^2}{2} - \frac{u_1^2}{2}\right)$$

or

$$W = \frac{mu_2^2}{2g_c} - \frac{mu_1^2}{2g_c} = \Delta\left(\frac{mu^2}{2g_c}\right)$$

where $(mu^2)/(2g_c)$ is called kinetic energy, or

$$E_K = \frac{mu^2}{2g_c} = \text{ft-lb}_f$$

Raising a mass m from an elevation Z_1 to an elevation Z_2 requires an upward force at least equal to the weight of the body and this force must move through the distance $Z_2 - Z_1$, or

$$F = ma/g_c = mg/g_c$$

Then

$$W = F(Z_2 - Z_1) = m(g/g_c)(Z_2 - Z_1)$$

$$= mZ_2\frac{g}{g_c} - mZ_1\frac{g}{g_c} = \Delta\left(\frac{mZg}{g_c}\right) = \Delta E_p$$

where E_p is called potential energy.

If a body is given energy when it is elevated, it is reasonable to assume that the body will retain this energy until it performs the work of which it is capable. So if an elevated body falls freely, it should gain in kinetic energy what it loses in potential energy, or

$$\Delta E_K + \Delta E_p = 0$$

This has been confirmed experimentally. So work is energy in transit.

5. Heat

When a hot body is brought into contact with a cold body, the hot body becomes cooler, the cooler body hotter. This "something" that is transferred from the hot body to the cold is called heat Q. Heat is a form of energy, measured as the calorie or the British thermal unit. These units are related to the primary energy unit, the joule, by

$$1 \text{ calorie} = 4.1840 \text{ J}$$

$$1 \text{ Btu} \quad = 1055.04 \text{ J}$$

and

$$1 \text{ Btu} \cong 1055 \text{ } J \cong 252 \text{ cal} \cong 778 \text{ ft-lb}_f$$

II. THE FIRST LAW AND SOME BASIC PRINCIPLES

A. Internal Energy

Energy added to a body as work exists as internal energy until it is extracted as heat. This internal energy U increases the energy of the molecules of the body.

B.   First Law

The total quantity of energy is constant. When it disappears in one form, it appears simultaneously in other forms. To use this concept, it is convenient to divide the world into two parts, the system and its surroundings. Everything not included in the system is considered to constitute the surroundings. And so the first law applies to the system and the surroundings, or

$$\Delta(\text{energy of the system}) + \Delta(\text{energy of surroundings}) = 0$$

If a system boundary does not allow the transfer of matter between the system and its surroundings, the system is *closed* and its mass is constant. Therefore all energy passing across the boundary between the system and its surroundings is transferred as heat and work. So for a closed system,

$$\Delta(\text{energy of surroundings}) = \pm Q \pm W,$$

with the sign choices depending upon which transfer direction is called positive. For the closed system,

$$\Delta(\text{energy of system}) = \Delta U + \Delta E_K + \Delta E_p$$

If *heat* is *positive* when transferred *to* the system from the *surroundings* and if *work* is *positive* when transferred *from* the *system to* the *surroundings*,

$$\Delta U + \Delta E_K + \Delta E_p = Q - W$$

or the total energy change of the system is equal to the heat added to the system less the work done by the system.

C.   State Functions

Thermodynamics is made up of two types of quantities: those dependent on path and those not. Those quantities not depending upon path are called *state* or *point* functions. Internal energy is a state function; work and heat, being path dependent, are not state functions.

D.   Enthalpy

Enthalpy is defined by the equation

$$H = U + PV$$

It is also a state function.

E.   The Steady-State Flow Process

Consider the process shown in Figure T2, with a heat exchanger and a turbine, operating at steady state (as opposed to the prior closed system). Our first-law equation for this system is

$$\Delta U + \frac{\Delta u^2}{2g_c} + \frac{g\,\Delta Z}{g_c} = Q - W$$

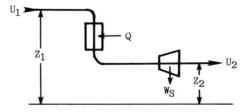

FIGURE T2.   Steady-state flow process.

but the W includes all work interchanged between the system and the surroundings, or

$$W = W_S + P_2 V_2 - P_1 V_1$$

Then

$$\Delta U + P_2 V_2 - P_1 V_1 + \frac{\Delta u^2}{2g_c} + \frac{g}{g_c} \Delta Z = Q - W_S$$

or

$$\Delta H + \frac{\Delta u^2}{2g_c} + \frac{g}{g_c} \Delta Z = Q - W_S$$

since

$$\Delta U + \Delta(PV) = \Delta H$$

Of course, for systems where kinetic and potential energy changes are negligible,

$$\Delta H = Q - W_S$$

F.   Equilibrium

An equilibrium system is one in which there is no tendency for a change in state to occur. There is no driving force present.

G.   The Phase Rule

The state of a pure homogeneous substance is fixed when two state functions are set. For more complex systems, this number changes. For nonreacting systems, J. W. Gibbs showed that the number of degrees of freedom F is given by

$$F = 2 - \pi - N$$

where π is the number of phases and N is the number of species. (This is sometimes remembered by $P + F = C + 2$, Police + Force = Chief + Two, $P =$ number of phases.)

H. Reversible Process

A frictionless process operating in a differential manner producing the maximum attainable work is a reversible process.

I. Specific Heat and Heat Capacity

A definite amount of heat is required to raise the temperature of a given mass of any material by one degree. The quantity is called the heat capacity of the substance. This relationship may be expressed as

$$dQ = nC \, dT$$

where n is the number of moles, C is the molar heat capacity, and dT is the rise in temperature caused by the dQ quantity of heat. Had we expressed C as the heat capacity per unit mass, we would have used the *specific heat.*

For a constant volume process

$$dU = dQ = C_v \, dT$$

or

$$\Delta U = Q = \int C_v \, dT$$

while for a constant pressure process

$$dH = C_p \, dT = dQ$$

or

$$\Delta H = \int C_p \, dT = Q$$

III. PURE FLUID VOLUMETRIC PROPERTIES

A. *PVT* Behavior

Homogeneous fluids are either gases or liquids, as shown in Figure T3. But the distinction becomes blurred at the critical point, where there is but one phase. Figure T3 provides no information about volume. If we were to draw isotherms on Figure T3 to the right of the solid region and plot pressure versus molar or specific volume for each isotherm, we would obtain Figure T4. The T_1 and T_2 lines are isotherms at temperatures greater than the critical temperature; they are smooth. This figure shows that where a single phase exists, there is a relation connecting P, V, and T, or $f(P, V, T) = 0$, an equation of state. Of course, the simplest is the ideal-gas law

$$PV = RT$$

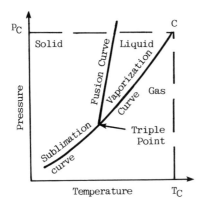

FIGURE T3. Pure fluid PT diagram.

B. The Virial Equation

Experiment has shown that we can write

$$PV = a(1 + B'P + C'P^2 + \cdots)$$

and when data for pure substances are plotted in this way and extrapolated back to $P = 0$, all substances intercept at the same point. So we fix the absolute temperature scale so that a is directly proportional to T, that is,

$$a = RT$$

where $R = 82.05$ cm^3 atm/g mol K, the universal gas constant. We can now rewrite our equation

$$Z = \frac{PV}{RT} = 1 + B'P + C'P^2 + \cdots$$

where Z is called the compressibility factor.

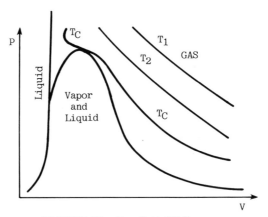

FIGURE T4. Pure fluid PV diagram.

Alternative we could write

$$Z = 1 + \frac{B}{V} + \frac{C}{V^2} + \cdots$$

Both equations are virial expansions; B and B' are called second virial coefficients, C and C' third virial coefficients, and so forth. Note that the two sets of coefficients are related by

$$B' = \frac{B}{RT}, \qquad C' = \frac{C - B^2}{(RT)^2}, \qquad D' = \frac{D - 3\,BC + 2\,B^3}{(RT)^3}, \qquad \text{etc.}$$

C. An Ideal Gas

The definition of an ideal gas is one whose equation of state is $Z = 1$ and whose internal energy is independent of pressure and volume, being a function of temperature only. If we apply the first law to reversible nonflow processes involving 1 mol of gas, then for

1. Constant-Volume (Isometric) Processes

$$dU = dQ = C_v\,dT$$

2. Constant-Pressure (Isobaric) Processes

$$dH = dU + R\,dT$$

or

$$C_p\,dT = C_v\,dT + R\,dT$$

or

$$C_p = C_v + R$$

3. Constant-Temperature (Isothermal) Processes

$$dU = dQ - dW = 0 \qquad \text{or} \qquad Q = W$$

$$Q = W = \int P\,dV = \int RT\,\frac{dV}{V}$$

or

$$Q = W = RT \ln \frac{V_2}{V_1} = RT \ln \frac{P_1}{P_2}$$

4.   Adiabatic Processes (dQ = 0)

$$dU = -dW = -P \, dV = -(RT/V) \, dV = C_v \, dT$$

or

$$\frac{dT}{T} = -\frac{R}{C_v}\frac{dV}{V}$$

If we define $\gamma = C_p/C_v$, then

$$\ln \frac{T_2}{T_1} = -(\gamma - 1) \ln \frac{V_2}{V_1}$$

or

$$\frac{T_2}{T_1} = \left(\frac{V_1}{V_2}\right)^{\gamma - 1}$$

The expression for work W becomes

$$W = \frac{P_1 V_1}{\gamma - 1}\left[1 - \left(\frac{P_2}{P_1}\right)^{(\gamma - 1)/\gamma}\right] = \frac{RT}{\gamma - 1}\left[1 - \left(\frac{P_2}{P_1}\right)^{(\gamma - 1)/\gamma}\right]$$

These expressions yield fairly good results for real gases. For monatomic gases, $\gamma = 1.67$; for diatomic gases, $\gamma = 1.40$; and for simple polyatomic gases (CO_2, SO_2, NH_3, and CH_4), $\gamma = 1.3$.

5.   Polytropic Processes

For this general case, when only the reversibility of an ideal gas is assumed:

$$dU = dQ - dW$$
$$dW = P \, dV$$
$$dU = C_v \, dT$$
$$dH = C_p \, dT$$

and so

$$dQ = C_p \, dT + P \, dV$$

D.   Cubic Equations of State

1.   van der Waals Equation

$$\left(P + \frac{a}{V^2}\right)(V - b) = RT, \qquad a = \frac{27 \, R^2 T_c^2}{64 P_c}, \qquad b = \frac{RT_c}{8P_c}$$

using

$$\left(\frac{\partial P}{\partial V}\right)_{T_c} = \left(\frac{\partial^2 P}{\partial V^2}\right)_{T_c} = 0$$

2. Redlich–Kwong Equation

$$P = \frac{RT}{V-b} - \frac{a}{T^{1/2}(V+b)V}$$

or

$$Z = \frac{1}{1-h} - \frac{A}{B}\left(\frac{h}{1+h}\right), \qquad h = \frac{b}{V} = \frac{BP}{Z}$$

$$B = \frac{b}{RT}, \qquad \frac{A}{B} = \frac{a}{bRT^{1.5}}$$

$$a = \frac{0.4278R^2T_c^{2.5}}{P_c}, \qquad b = \frac{0.0867\,RT_c}{P_c}$$

E.   Generalized Correlations and Acentric Factor

The principle of corresponding states (using reduced temperatures and reduced pressures) is much better than the ideal-gas law. To make the results even better. Pitzer developed a third factor, the acentric factor ω, determined from the vapor pressure measured at a reduced temperature of 0.7,

$$\omega = -\log_{10}(P_r^{\text{sat}})_{T_r=0.7} - 1.000$$

where $P_r = P/P_c$, $T_r = T/T_c$.
    Based upon this idea

$$Z = 1 + \left(\frac{BP_c}{RT_c}\right)\left(\frac{P_r}{T_r}\right)$$

$$\frac{BP_c}{RT_c} = B^0 + \omega B^1$$

$$B^0 = 0.083 - \frac{0.422}{T_r^{1.6}}$$

$$B^1 = 0.139 - \frac{0.172}{T_r^{4.2}}$$

This equation is accurate for $T_r > 0.486 + 0.457\,P_r$.
    For conditions below this line, use

$$Z = Z^0 + \omega Z^1$$

where Z^0 and Z^1 are obtained from generalized curves such as those given by Smith and Van Ness, pp. 89–90.

F. Liquid Behavior

The Tait equation

$$V = V_0 - D \ln\left(\frac{P + E}{P_0 + E}\right)$$

gives excellent results for liquids. Here V_0, P_0 are a set of reference state volume and pressure, and D, E are correlated from experimental data at a given temperature

IV. THE SECOND LAW OF THERMODYNAMICS

A. Statements of the Second Law

There are many different ways of stating the second law of thermodynamics. Some of them are

1. Any process that consists solely in the transfer of heat from one temperature to a higher one is impossible.
2. Nothing can operate in such a way that its only effect (in both system and surroundings) is to convert heat absorbed by a system completely into work.
3. It is impossible to convert the heat absorbed completely into work in a cyclical process.

Thus the second law-does not prevent the production of work from heat, but it does limit the efficiency of any cyclical process.

B. The Heat Engine

The heat engine, a machine that produces work from heat in a cyclical process, illustrates the second law. For instance, a steam plant operates as follows:

1. Part of the heat from the fuel is transferred to liquid water in the boiler, converting it into steam at a high pressure and temperature. Let the heat absorbed by the steam be Q_1.
2. Energy is transferred from the steam as shaft work by a turbine. This is adiabatic.
3. Exhaust steam from the turbine is condensed at a low temperature by the transfer of heat to cooling water. Call this Q_2. *Note that, for the time being, the numerical values of Q_1 and Q_2 are regarded as positive.*
4. Liquid water is pumped back to the boiler, completing the cycle. This is adiabatic.

Now the system energy change must be zero for the complete cycle, or

$$\Delta U_{\text{cycle}} = 0 = Q_1 - Q_2 - W$$

or

$$Q_1 - Q_2 = W$$

The ratio of W to Q_1, the efficiency of the engine for converting heat into work, would be expected to depend upon the degree of reversibility of processes 1–4. In fact, an irreversible heat engine cannot have an efficiency greater than a reversible one.

C. The Ideal-Gas Temperature Scale

Consider an ideal gas as the working fluid in a reversible cyclical heat engine or Carnot engine as illustrated in Figure T5. The cycle consists of an isothermal expansion from A to B, during which an amount of heat Q_1 is absorbed at T_1. Then we have an adiabatic expansion BC to a lower temperature T_2. Following this, we discard an amount of heat Q_2 at the constant temperature T_2 while undergoing the isothermal compression from C to D.

Finally the gas returns adiabatically to its original state by compression D to A.

Now the net work done by the gas is the algebraic sum of the work effects for each of the steps:

$$W_{\text{net}} = W_{\text{ab}} + W_{\text{bc}} + W_{\text{cd}} + W_{\text{da}}$$

The BC and DA processes are reversible and adiabatic, so the first law says that the work is equal to $-\Delta U$,

$$W_{\text{bc}} = -\Delta U_{\text{bc}} = -\int_{T_1}^{T_2} C_v \, dT$$

$$W_{\text{da}} = -\Delta U_{\text{da}} = -\int_{T_2}^{T_1} C_v \, dT$$

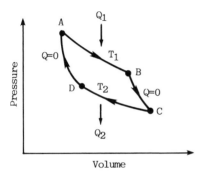

FIGURE T5.   Carnot cycle PV diagram.

for each mole. Thus these two amounts of work cancel each other and

$$W_{net} = W_{ab} + W_{cd}$$

But these are isothermal processes, so

$$W_{net} = RT_1 \ln\left(\frac{P_a}{P_b}\right) + RT_2 \ln\left(\frac{P_c}{P_d}\right)$$

and for isothermal processes $\Delta U = 0$ so

$$Q_1 = W_{ab} = RT_1 \ln(P_a/P_b)$$

The efficiency, or the ratio of work obtained to the heat absorbed, W_{net}/Q_1, is given by

$$\frac{W_{net}}{Q_1} = \frac{1}{T_1}\left\{T_1 + T_2\left[\frac{\ln(P_c/P_d)}{\ln(P_a/P_b)}\right]\right\}$$

Because $PV = RT$ for all the processes in this ideal cycle, and $PV^\gamma = $ constant for adiabatic reversible processes,

$$\frac{P_b}{P_a} = \frac{P_c}{P_d}$$

Then

$$\frac{W_{net}}{Q_1} = \frac{T_1 - T_2}{T_1}$$

or, since $Q_2 = Q_1 - W_{net}$,

$$\frac{Q_2}{Q_1} = \frac{T_2}{T_1}$$

D. Entropy

We now return to the usual sign convention that heat is positive when absorbed by the system and negative when rejected by the system. The prior equation then is really

$$\frac{Q_1}{T_1} = -\frac{Q_2}{T_2}$$

or

$$\frac{Q_1}{T_1} + \frac{Q_2}{T_2} = 0$$

This suggests that the quantities Q_1/T_1 and Q_2/T_2 represent property changes of the working fluid because their sum is zero for the cycle, a characteristic of a property or state function. Expressed differentially,

$$\frac{dQ_1}{T_1} + \frac{dQ_2}{T_2} = 0$$

The integral of dQ_R/T (R emphasizes the necessity of reversibility) is the same for any reversible path and so depends only upon the initial and final states. This quantity has been given the name *change in entropy* ΔS. The calculation of entropy changes must always be for reversible conditions.

Entropy ideas may be summarized:

1. The entropy change of a system undergoing a reversible process is equal to $dS = dQ_R/T$ or

$$\Delta S = \int \frac{dQ_R}{T}$$

2. To calculate the ΔS for a system undergoing an irreversible change in state, we must devise a reversible process for accomplishing the same change and then evaluate $\int dQ/T$ over the reversible process. This will give us the entropy change for the original irreversible process.

3. If *a process involves only the transfer of energy as heat to a system*, the entropy change is correctly evaluated from $\int dQ/T$ for the actual process, even if it is irreversible.

E. Real Processes and the Second Law

All processes proceed in a direction such that the total entropy change is positive, approaching a limit of zero, as the process approaches reversibility, or

$$\Delta S_{total} \geq 0$$

No process can be invented for which the total entropy decreases.

F. Irreversibility and Entropy Change

We develop the relationship between the total entropy change and the degree of irreversibility with the following example.

When an amount of heat Q is transferred from a system at a temperature T_1 to surroundings at the lower temperature T_2, the total entropy change is given by

$$\Delta S_{total} = \Delta S_{surr} + \Delta S_{sys} = \frac{Q}{T_2} - \frac{Q}{T_1} = Q\left(\frac{T_1 - T_2}{T_1 T_2}\right)$$

On the other hand, suppose that the heat Q is transferred to a reversible heat engine that absorbs this amount at T_1, converting part to work and rejecting the remainder at T_2. The quantity of work that could be obtained, but which was lost because of the irreversible nature of the actual process is given by

$$W = Q\left(\frac{T_1 - T_2}{T_1}\right) = T_2 Q\left(\frac{T_1 - T_2}{T_1 T_2}\right)$$

So if we eliminate Q between these equations,

$$W_{lost} = T_2(\Delta S_{surr} + \Delta S_{sys}) = T_2 \, \Delta S_{total}$$

G. The Third Law of Thermodynamics

The absolute entropy is zero for all perfect crystalline substances at absolute zero temperature. Thus, the absolute entropy of a gas at temperature T is given by

$$S = \int_0^{T_f} \frac{(C_p)_s}{T} \, dT + \frac{\Delta H_f}{T_f} + \int_{T_f}^{T_v} \frac{(C_p)_l}{T} \, dT + \frac{\Delta H_v}{T_v} + \int_{T_v}^{T} \frac{(C_p)_g}{T} \, dT$$

where T_f, T_v = temperature of fusion, vaporization and

ΔH_f, ΔH_v = heat of fusion, vaporization.

V. THERMODYNAMIC PROPERTIES OF FLUIDS

A. Some Common Thermodynamic Definitions

U is the internal energy of a system; it includes all forms of energy contained in a system except the energy due to position (potential) or motion (kinetic) of the system.

$H = U + PV$ is the enthalpy. It is purely a convenient thermodynamic function relating internal energy and the product of pressure and volume.

$A = U - TS$, the Helmholtz free energy, is a convenient thermodynamic function relating internal energy and the product of the two fundamental properties, temperature and entropy.

$G = H - TS$ is another convenient thermodynamic function like A. It is most frequently referred to as the Gibbs free energy, although there are those that say this is a misnomer.

Anyone who has studied thermodynamics is appalled at trying to learn all the interrelationships between the various thermodynamic terms. We shall use very few of them, but for general information a memorization method is presented.

B. Thermodynamic Relationships

The method consists of eight thermodynamic functions arranged in a square.[†] The order of the symbols of the functions can be remembered by means of

† Method of Shih-Ching Su, modified by Chin-Yung Wen.

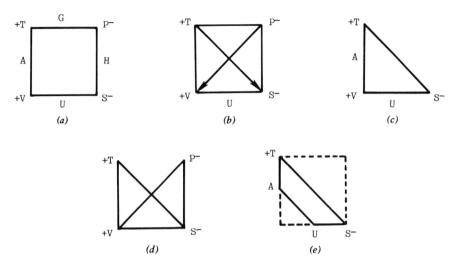

FIGURE T6. Thermodynamic relationships.

the coded sentence; "The Gods Punish Humble Souls Until Vanquished Altogether," with the initial letters placed as shown in Figure T6a.

1. To obtain dU as a function of the appropriate independent variables, draw the two diagonals toward U, using the signs corresponding to the symbols at the base of the vectors, as shown in Figure T6b, to get

$$dU = T\,dS - P\,dV$$

In a like manner,

$$dA = -S\,dT - P\,dV$$
$$dG = V\,dP - S\,dT$$
$$dH = V\,dP + T\,dS$$

2. The partial derivatives may be obtained from triangular relationships, as shown in Figure T6c,

$$\frac{(\partial U)}{(\partial S)_V} = T$$

where the variable outside the parentheses, V in this case, is always at the right angle. From this same triangle, we obtain

$$\frac{(\partial A)}{(\partial T)_V} = -S$$

Using the other triangles, we obtain

$$\frac{(\partial U)}{(\partial V)_S} = -P, \quad \frac{(\partial H)}{(\partial P)_S} = V$$

$$\frac{(\partial H)}{(\partial S)_P} = T, \quad \frac{(\partial G)}{(\partial T)_P} = -S$$

$$\frac{(\partial G)}{(\partial P)_T} = V, \quad \frac{(\partial A)}{(\partial V)_T} = -P$$

3. The Maxwell relationships can be obtained by considering the right triangle with the same base, as shown in Figure T6d,

$$\frac{(\partial T)}{(\partial V)_S} = -\frac{(\partial P)}{(\partial S)_V}$$

Likewise with the others

$$\frac{(\partial V)}{(\partial S)_P} = \frac{(\partial T)}{(\partial P)_S}, \quad -\frac{(\partial S)}{(\partial P)_T} = \frac{(\partial V)}{(\partial T)_P}, \quad -\frac{(\partial P)}{(\partial T)_V} = -\frac{(\partial S)}{(\partial V)_T}$$

4. Four other relationships may be derived from the corners of Figure T6a:

$$\frac{(\partial T)}{(\partial V)_A} = -\frac{P}{S}, \quad \frac{(\partial V)}{(\partial S)_U} = \frac{T}{P}$$

$$\frac{(\partial S)}{(\partial P)_H} = \frac{V}{T}, \quad -\frac{(\partial P)}{(\partial T)_G} = -\frac{S}{V}$$

5. Another four relationships may be obtained from the trapezoids, as shown in Figure T6e,

$$U - A = TS$$

Likewise,

$$A - G = -PV, \quad G - H = -ST, \quad H - U = VP$$

6. Many other relationships, such as chemical potentials, also can be derived, for example, by considering each side of Figure T6a,

$$\mu = \frac{(\partial U)}{(\partial n)_{S,V}} = \frac{(\partial A)}{(\partial n)_{T,V}} = \frac{(\partial G)}{(\partial n)_{P,T}} = \frac{(\partial H)}{(\partial n)_{P,S}}$$

VI. PROPERTIES OF HOMOGENEOUS MIXTURES

A. Fugacity

Gibbs free energy G is defined by

$$G = H - TS$$

It has a unique relation to temperature and pressure since $dH = T\,dS + V\,dP$, for if we differentiate Gibbs free energy

$$dG = dH - T\,dS - S\,dT$$
$$= -S\,dT + V\,dP$$

Applying this to 1 mol of a pure fluid i at constant T,

$$dG_i = V_i\,dP$$

But for an ideal gas $V_i = RT/P$, so

$$dG_i = RT\,\frac{dP}{P}$$

or

$$dG_i = RT\,d\ln P$$

(at constant T). This suggests replacing the pressure P with a new function, f_i, or the fugacity of pure i,

$$dG_i = RT\,d\ln f_i$$

with f_i having the units of pressure. To complete the definition,

$$\lim_{P \to 0} \frac{f_i}{P} = 1$$

The fugacity of a component in solution, $\hat{f}_i$, is defined likewise:

$$d\bar{G}_i = RT\,d\ln \hat{f}_i, \qquad \lim_{P \to 0} \frac{\hat{f}_i}{x_i P} = 1$$

(at constant T). Thus for an ideal gas

$$\hat{f}_i = x_i P$$

and this is the definition of partial pressure.

B.   Fugacity Coefficient

The fugacity coefficient is the ratio of the fugacity of a material to its pressure. For a pure substance, $\phi_i = f_i/P$. For a component in solution, $\phi_i = f_i/(x_i P)$. These values may be calculated from PVT data.

$$dG_i = V_i\,dP = RT\,d\ln f_i$$

(at constant T). Now since

$$f_i = P\phi_i, \qquad \ln f_i = \ln P + \ln \phi_i, \qquad \text{and } d\ln f_i = \frac{dP}{P} + d\ln \phi_i$$

and

$$V_i \, dP = RT\left(\frac{dP}{P} + d \ln \phi_i\right)$$

or

$$d \ln \phi_i = \left(\frac{PV_i}{RT}\right)\left(\frac{dP}{P}\right) - \frac{dP}{P}$$

But $Z = PV_i/(RT)$, so

$$d \ln \phi_0 = (Z_i - 1) \, dP/P$$

or

$$\int_{\phi_i=1}^{\phi_i} d \ln \phi_i = \ln \phi_i = \int_0^P (Z_i - 1) \frac{dP}{P}$$

(at constant T). Equations and generalized plots have been developed from this equation (see Smith and Van Ness, pp. 236–237).

C. Activity

The activity $\hat{a}_i$ of component i in solution is defined to be

$$\hat{a}_i = \frac{\hat{f}_i}{f_i^\circ}$$

where f_i° is the fugacity of component i in a standard state at the same temperature and pressure as that of the mixture. Now if

$$\frac{\Delta G}{RT} = \frac{1}{RT} \sum x_i(\bar{G}_i - G_i^\circ) = \sum (x_i \ln \hat{a}_i)$$

For an ideal solution,

$$\left(\frac{\Delta G}{RT}\right)_{\text{ideal}} = \sum x_i \ln x_i$$

D. Activity Coefficients and Excess Properties

An excess property is the difference between an actual property and the property that would be calculated for the same conditions of T, P, and x by the equations of an ideal solution. Superscript E denotes the excess property. Using this concept, it is possible to obtain

$$\frac{G^E}{RT} = \sum (x_i \ln \gamma_i)$$

where the activity coefficient γ_i is given by

$$\gamma_i = \frac{\hat{a}_i}{x_i} = \frac{\hat{f}_i}{x_i f_i^\circ}$$

Based upon the partial molar excess property,

$$\frac{\bar{G}_i^{\mathrm{E}}}{RT} = \ln \gamma_i$$

VII. PHASE EQUILIBRIA

A. Criteria

The equilibrium state of a closed system is that state for which the total Gibbs free energy is a minimum with respect to all possible changes at the given T and P, that is, $(dG)_{T,P} = 0$.

B. Vapor–Liquid Equilibrium

Using the fugacity coefficient for each phase and equating them at equilibrium,

$$y_i \hat{\phi}_i^\mathrm{v} = x_i \hat{\phi}_i^\mathrm{l}$$

but the $\hat{\phi}_i^\mathrm{v}$ and ϕ_i^l are composition dependent. When the vapor phase is an ideal gas and the liquid phase is an ideal solution, we can reduce this equation to Raoult's law:

$$y_i = x_i \left(\frac{P_i^{\mathrm{sat}}}{P} \right)$$

C. Activity Coefficients from Experimental Data

Activity coefficients for a given system are related to G^{E}/RT, so most correlations are based upon this function. The excess Gibbs free energy G^{E} is a function of T, P, and the x_i's but for liquids the P effect is weak. So for low pressure at constant T

$$\frac{G^{\mathrm{E}}}{RT} = g(x_1, x_2, \ldots, x_N)$$

A power series, the Redlich–Kister expansion,

$$\frac{G^{\mathrm{E}}/RT}{x_1 x_2} = B + C(x_1 - x_2) + D(x_1 - x_2)^2 + \cdots$$

is effective for binary mixtures. For instance, if we use

$$\frac{G^E/RT}{x_1 x_2} = B$$

the corresponding equations for $\ln \gamma_1$ and $\ln \gamma_2$ are

$$\ln \gamma_1 = B x_2^2, \ln \gamma_2 = B x_1^2$$

If we use

$$\frac{G^E/RT}{x_1 x_2} = B + C(x_1 - x_2)$$

then

$$\ln \gamma_1 = x_2^2 [A_{12} + 2(A_{21} - A_{12})x_1]$$
$$\ln \gamma_2 = x_1^2 [A_{21} + 2(A_{12} - A_{21})x_2]$$

the Margules equations. If we use

$$\frac{x_1 x_2}{G^E RT} = B' + C'(x_1 - x_2)$$

then we obtain

$$\ln \gamma_1 = x_2^2 A'_{12} \bigg/ \left[1 + \left(\frac{A'_{12}}{A'_{21}} - 1 \right)x_1 \right]^2 = A'_{12} \bigg/ \left(1 + \frac{A'_{12} x_1}{A'_{21} x_2} \right)^2$$

$$\ln \gamma_2 = x_1^2 A'_{12} \bigg/ \left[1 + \left(\frac{A'_{21}}{A'_{12}} - 1 \right)x_2 \right]^2 = A'_{21} \bigg/ \left(1 + \frac{A'_{21} x_1}{A'_{12} x_1} \right)^2$$

the van Laar equations.

VIII. CHEMICAL REACTION EQUILIBRIUM

A. The Reaction Coordinate

Following Aris and Petersen, if we write the stoichiometric equation so that the stoichiometric numbers v_i are positive for a product and negative for a reactant, then we may write

$$\frac{dn_1}{v_1} = \frac{dn_2}{v_2} = \cdots = d\varepsilon$$

or

$$dn_i = v_i\, d\varepsilon$$

where n_i = moles of species i and ε = reaction coordinate or extent of reaction.

When two or more reactions proceed simultaneously, a separate reaction coordinate ε_j is associated with each reaction j. If we have r independent reactions, and the stoichiometric numbers of the species A_i are denoted for each species and each reaction by $v_{i,j}$ ($i = 1, 2, \ldots, N$ are species and $j = 1, 2, \ldots, r$ are the reactions), then

$$\sum_i v_{i,j} A_i = 0 \qquad (j = 1, 2, \ldots, r)$$

and

$$dn_i = \sum_j v_{i,j} d\varepsilon_j \qquad (i = 1, 2, \ldots, N)$$

B. Equilibrium and Gibbs Free Energy

The total differential of the Gibbs free energy of a single-phase, multicomponent system is

$$dG^t = -S^t \, dT + V^t \, dP + \sum \mu_i \, dn_i$$

or

$$dG^t = -S^t \, dT + V^t \, dP + \sum (\mu_i v_i) \, d\varepsilon$$

where μ_i is the chemical potential of component i. Note that from this we obtain

$$\sum (\mu_i v_i) = \left(\frac{\partial G^t}{\partial \varepsilon} \right)_{T,P}$$

and, at equilibrium,

$$\sum \mu_i v_i = 0$$

Now

$$\mu_i = G_i^\circ + RT \ln \hat{a}_i$$

So

$$\sum v_i G_i^\circ + RT \sum \ln a_i^{v_i} = 0$$

or

$$\ln \prod a_i^{v_i} = -\sum v_i G_i^\circ / RT$$

where $\prod$ means the product over all i. It is usual to define

$$K = \prod a_i^{v_i}$$

So, finally,

$$-RT \ln K = \sum v_i G_i^\circ = \Delta G^\circ$$

where K is the equilibrium constant and ΔG° is the standard Gibbs free energy change of reaction. The activities $\hat{a}_i = \hat{f}_i/f_i^\circ$ are the connection between the equilibrium state and the standard states of the constituents at the equilibrium temperature T.

Sometimes data are given such that one must use

$$\Delta G^\circ = \Delta H^\circ - T\Delta S^\circ$$

where ΔH° and ΔS° are the standard heat of reaction and the standard entropy change of reaction.

C. Temperature Effect

The van't Hoff equation relates the equilibrium constant to the temperature

$$\frac{d \ln K}{dT} = \frac{\Delta H^\circ}{RT^2}$$

D. Equilibrium Constant and Composition

For gas-phase reactions

$$K = \prod f_i^{\nu_i}$$

where the fugacities are functions of T, P, and composition. But K is a function of temperature only. So for a fixed temperature, the composition at equilibrium must change with pressure in a way that $\prod f_i^{\nu_i}$ remains constant. Now

$$f_i = \hat{\phi}_i y_i P$$

with $\hat{\phi}_i$ the fugacity coefficient, y_i the mole fraction of component i in the equilibrium mixture, and P the equilibrium pressure. So

$$\prod (y_i \hat{\phi}_i)^{\nu_i} = P^{-\nu} K$$

where $\nu = \sum \nu_i$ and P is in atmospheres. The y_i's may be expressed in terms of the reaction coordinate at equilibrium ε_e. The $\hat{\phi}_i$ may depend on composition and thus upon ε_e.

If the equilibrium mixture is an ideal solution, then $\hat{\phi}_i = \phi_i$, the fugacity coefficient of pure i at T and P. For this case

$$\prod (y_i \phi_i)^{\nu_i} = P^{-\nu} K$$

and the ϕ_i may be evaluated from generalized correlations if T and P are known.

If the temperature is sufficiently high or the pressure is sufficiently low, the equilibrium mixture will be an ideal gas and $\hat{\phi}_i = 1$ or

$$\prod y_i^{\nu_i} = P^{-\nu} K$$

IX. HEAT INTO WORK BY POWER CYCLES

A. Behavior of Some Heat Engines

Several of the more important heat engines will be discussed here. A much larger selection may be found in Reynolds and Perkins, *Engineering Thermodynamics*, McGraw-Hill, New York, 1970.

The steps in the cycle of each engine are shown in Table T1. The engine performance can be obtained by solving for the heat added and the work done during each step. The overall performance can be calculated using the given efficiency equations, based upon the nomenclature of Figure T7. To illustrate

TABLE T1[a]

Carnot Cycle	Otto Cycle (air standard)	Diesel Cycle (air standard)	Gas Turbine (air standard)
$\eta = \dfrac{T_1 - T_2}{T_1}$	$\eta = 1 - \left(\dfrac{1}{r}\right)^{\gamma - 1}$	$\eta = 1 - \dfrac{\left(\dfrac{1}{r_e}\right)^{\gamma} - \left(\dfrac{1}{r_c}\right)^{\gamma}}{\gamma\left(\dfrac{1}{r_e} - \dfrac{1}{r_c}\right)}$	$\eta = 1 - \left(\dfrac{P_a}{P_b}\right)^{(\gamma - 1)/\gamma}$
Adiabatic compression	Adiabatic compression	Adiabatic compression	Isentropic compression
Isothermal expansion	Const. volume heat absorp.	Const. pressure expansion	Const. pressure heat absorption
Adiabatic expansion	Adiabatic expansion	Adiabatic expansion	Isentropic expansion
Isotherm compression	Const. volume heat rejection	Const. volume heat rejection	Const. pressure heat rejection

[a] Notes

$T_1 > T_2$

r = compression ratio $= \dfrac{V \text{ before comp.}}{V \text{ after comp.}}$

r_c = compression ratio

$r_e = \dfrac{\text{expansion}}{\text{ratio}} = \dfrac{V \text{ after expan.}}{V \text{ before expan.}}$

$P_b > P_a$

$\gamma = C_p/C_v$

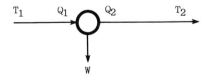

T₁   Q₁   Q₂   T₂

W

FIGURE T7. Nomenclature for heat engines of Table T1.

the use of Table T1, consider this problem: A Carnot engine and air-standard Otto, diesel, and gas turbine engines are each operating with a heat addition rate of 10,000 Btu/hr. The Carnot engine is operating between heat reservoirs of 1200 and 300°F. The other engines have compression ratios of eight to one, and the diesel engine has an expansion ratio of two to one. Find the net horsepower produced by each engine.

Solution: For the *Carnot* engine

$$\eta = \frac{T_1 - T_2}{T_1} = \frac{1200 - 300}{1200 + 460} = 0.5422$$

$$W = Q_1\eta \qquad = (10,000 \text{ Btu/hr})(0.5422)$$

$$= 5422 \text{ Btu/hr}$$

Converting to horsepower,

$$W = \frac{(5422 \text{ Btu/hr})(778 \text{ ft lb}_\text{f}/\text{Btu})}{(550 \text{ ft lb}_\text{f}/\text{sec hp})(3600 \text{ sec/hr})} = 2.13 \text{ hp}$$

For the *Otto* engine, $r = 8$ and the efficiency is

$$\eta = 1 - (1/r)^{\gamma - 1} = 1 - (\tfrac{1}{8})^{0.4} = 0.5647 \quad \text{or} \quad 56.6\%$$
$$W = Q_1\eta = (10,000 \text{ Btu/hr})(0.5647) = 5647 \text{ Btu/hr} = 2.22 \text{ hp}$$

For the *diesel* engine, $r = 8, r_\text{e} = 2,$

$$\eta = 1 - \frac{r \, r_\text{e}^{-\gamma} - r^{-\gamma}}{\gamma \, (r/r_\text{e}) - 1} = 1 - \frac{8}{1.4}\left[\frac{2^{-1.4} - 8^{-1.4}}{(8/2) - 1}\right] = 0.3819$$

$$W = Q_1\eta = (10,000 \text{ Btu/hr})(0.3819) = 3819 \text{ Btu/hr} = 1.50 \text{ hp}$$

For the *gas turbine* engine, $r = 8,$

$$\eta = 1 - (1/r)^{(\gamma - 1)/\gamma} = 1 - (\tfrac{1}{8})^{0.4/1.4} = 0.4480$$
$$W = Q_1\eta = 4480 \text{ Btu/hr} = 1.76 \text{ hp}$$

B.   *T–S* and *P–V* Diagrams for Representing Heat Engine Performance

In understanding the operation of a heat engine, it is often useful to plot the operation on a diagram of temperature versus entropy (T–S) or pressure versus volume (P–V). On a plot of pressure versus volume, each of the heat engine processes can be represented as a line as shown in Figure T8. The lines are

1. $V = $ constant, constant volume process;
2. $P = $ constant, constant pressure process;
3. $PV = $ constant, a process at constant temperature and enthalpy; and
4. $PV^\gamma = $ constant, a process at constant entropy (adiabatic and reversible).

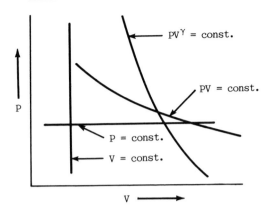

FIGURE T8. General *PV* diagram.

Now let us use the previous description of heat engines to plot a *P-V* diagram. For the Carnot cycle, as shown in Figure T9, we have:

a → b: adiabatic compression from P_a, V_a to P_b, V_b. Temperature rises from T_2 to T_1. Entropy remains the same.

b → c: heat adsorption at constant temperature T_1. Pressure decreases from P_b to P_c while volume increases from V_b to V_c. Enthalpy remains the same, while entropy increases.

c → d: adiabatic expansion from P_c, V_c to P_d, V_d. Temperature falls from T_1 to T_2. Entropy remains the same.

d → a: heat rejection at constant temperature T_2. Pressure increases from P_d to P_a while volume decreases from V_d to V_a. Enthalpy remains the same, and entropy decreases.

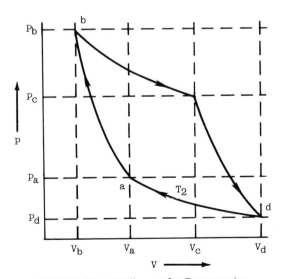

FIGURE T9. *PV* diagram for Carnot engine.

X. REFRIGERATION AND LIQUEFACTION

A. Carnot Refrigeration Cycle

For a continuous refrigeration process, since heat must be absorbed at a low temperature, there must be continuous rejection of heat to the surroundings at a higher temperature. So a refrigeration cycle is simply a reversed heat-engine cycle. Since heat is transferred from a low level to a higher one, external energy must be used so as not to violate the second law. The ideal refrigeration cycle is the Carnot: two isothermal processes where the heat Q_2 is absorbed at the lower T_2 temperature and heat Q_1 is rejected at the higher T_1 temperature; and two adiabatic processes whose result is the addition of the net work W to the system. Note that here we use both Q_1 and Q_2 positive so W is positive and equal to the work added to the system.

From the first law, since all ΔU of the fluid is zero for the whole cycle

$$W = Q_1 - Q_2$$

From the Carnot cycle of Figure T10,

$$Q_1 = T_1 \Delta S, \qquad Q_2 = T_2 \Delta S$$

so

$$\frac{W}{Q_2} = \frac{T_1 - T_2}{T_2}$$

We use this to calculate the required work W for a given quantity of refrigeration Q_2 using the ideal Carnot cycle.

B. Air-Refrigeration Cycle

The air-refrigeration cycle is not ideal; it can be built, but usually is not. In this cycle, the refrigerator air absorbs heat at an essentially constant pressure P_1 in the cold space, and, in the cooler, rejects heat to the surroundings at a higher, constant P_2. The gas is compressed (ideally) at constant entropy, A to B, using

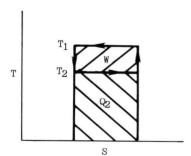

FIGURE T10. Idealized Carnot T–S diagram.

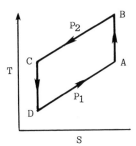

FIGURE T11. Air-refrigeration T–S diagram.

as part of the energy required, the work obtained from the expansion process CD. This process is illustrated in the T–S diagram of Figure T11.

The thermodynamic analysis of this cycle will illustrate the type of calculations needed for other, more practical cycles. If the specific heat of the gas is constant, the heat absorbed in the refrigerator, Q_2, and rejected in the cooler, Q_1, are

$$Q_2 = mC_p(T_A - T_D)$$

$$Q_1 = mC_p(T_B - T_C)$$

For the reversible adiabatic processes CD and AB, the P–T relationship is

$$\frac{T_C}{T_D} = \left(\frac{P_2}{P_1}\right)^{(\gamma-1)/\gamma} = \frac{T_B}{T_A}$$

Using $W = Q_1 - Q_2 = mC_p[(T_B - T_C) - (T_A - T_D)]$, the efficiency W/Q_2 is

$$\frac{W}{Q_2} = \frac{T_B}{T_A} - 1$$

Since refrigeration performance is defined in terms of a coefficient of performance (COP),

$$COP = \frac{\text{refrigeration obtained}}{\text{work required}} = \frac{Q_2}{W}$$

for this air cycle

$$COP = \frac{Q_2}{W} = \frac{T_A}{T_B - T_A}$$

C. Vapor-Compression Cycle

Constant-pressure evaporation of a liquid is a method of absorbing heat at constant temperature; constant-temperature condensation of a gas at a higher pressure is a method of rejecting heat. This scheme is the vapor-compression cycle:

$$\begin{array}{rl}
\text{constant pressure evaporation:} & \text{D} \rightarrow \text{A} \\
\text{isentropic compression:} & \text{A} \rightarrow \text{B} \\
\text{constant pressure condensation:} & \text{B} \rightarrow \text{C} \\
\text{adiabatic expansion:} & \text{C} \rightarrow \text{D}
\end{array}$$

This leads to

$$\text{COP} = \frac{H_A - H_D}{H_B - H_A}$$

$$\text{circulation rate} = m = \frac{12,000}{H_A - H_D}$$

$$= \text{lb per hr per ton of refrigeration}$$

Note: A ton of refrigeration = 12,000 Btu/hr.

D. Liquefaction Processes

The basis of all liquefaction processes is to cool the gas until it enters the two-phase region. It is done by three main methods:

1. Cooling at constant pressure (heat exchanger).
2. Cooling by expansion in an engine from which work is obtained.
3. Cooling by an expansion-valve or throttling process.

Of these, the *constant-enthalpy* or *Joule–Thomson expansion* (3) is the one most used.

PROBLEMS

A. Steam comes from a boiler at 650 psia and 760°F. After expansion to 150 psia, the steam is reheated to 760°F. Now expansion occurs to 3 in. Hg abs. For the Rankine cycle with reheating (ideal reheat cycle), calculate, for 1 lb of steam, (a) the heat added, (b) the heat rejected, (c) net work, (d) thermodynamic efficiency, and (e) quality of exhaust steam. Consider a Rankine cycle operating between the upper limit of 650 psia, 760°F, and the lower limit of 3 in. Hg abs. Compute these quantities. Compare the thermodynamic efficiencies of the two cycles. Also compare the thermodynamic efficiencies with the Carnot efficiency.

B. (a) Calculate the change in entropy when 1 lb of NH_3 is cooled from 400 to −175°F at a constant pressure of 1 atm. Express results as Btu/lb°R. (b) Calculate the absolute entropy of solid NH_3 at its melting point at 1-atm pressure, in Btu/lb°R. Boiling point = −33.4°C; melting point $= -77.7°C$; $C_p(\text{liq}) = 1.06$ cal/g K; $C_p(\text{solid}) = 0.502$ cal/g K; $\lambda_v = 5581$ cal/g mol; $\lambda_f = 1352$ cal/g mol; absolute entropy of gas at $25°C = 46.03$ cal/g mol K.

C. A hydrocarbon mixture (15 mol% methane, 15% ethane, 25% propane, 20% isobutane, 20% n-butane, and 5% n-pentane) is flashed into a separator at 80°F and 150 psia. Fifty two percent of the mixture leaves the separator as liquid. What is the composition of this liquid? K values: methane, 17.0; ethane, 3.1; propane, 1.0; isobutane, 0.44; n-butane, 0.32; n-pentane, 0.096.

D. Iron oxide (FeO) is reduced to Fe by passing a mixture of 30 mol% CO and 70 mol% N_2 over the oxide at 100 atm and 100°C. If $K_a = 35$ (at the stated conditions) for the reaction

$$\text{FeO(s)} + \text{CO(g)} \rightleftharpoons \text{Fe(s)} + \text{CO}_2(\text{g})$$

what is the weight of metallic Fe produced per 1000 ft^3 of the entering gas measured at 1 atm and 25°C?

Calculate the above if pure CO had been used instead of the mixture. Assume chemical equilibrium is reached in both cases.

E. A mixture (58.3 mol% n-hexane, 8.33% n-heptane, 33.3% steam) is cooled at a constant pressure (1 atm) from an initial temperature of 350°F. If we assume that the hydrocarbons and water are immiscible, calculate:

1. Temperature at which condensation first occurs.
2. Composition of the first liquid phase as it first appears.
3. Temperature at which the second liquid phase appears.
4. Composition of the second liquid phase as it first appears.

$\ln P = A + B/T$: $A = 17.7109$, $B = -6816.4$ for n-hexane; $A = 17.9184$, $B = -7547.4$ for n-heptane; $P = $ vapor pressure (mm Hg), $T = °R$.

F. It is claimed that pure ethane at 0.10 atm abs and 1000 K can be converted to ethylene and hydrogen and 93% conversion is claimed. Is this possible?

Ethane	$\Delta G°_{298} = -7860$ cal/g mol
(C_2H_6)	$\Delta H°_{298} = -20,236$ cal/g mol
	$C_p° = 2.3 + 0.02T$(cal/g mol K)
Ethylene	$\Delta G°_{298} = 16,282$ cal/g mol
(C_2H_4)	$\Delta H°_{298} = 12,500$ cal/g mol
	$C_p° = 2.8 + 0.03T$(cal/g mol K)
Hydrogen	$\Delta G°_{298} = \Delta H°_{298} = 0$
(H_2)	$C_p° = 6.9 + 0.004\ T$(cal/g mol K)

$G = H - TS = $ Gibbs free energy; $\Delta G°_{298}$, $\Delta H°_{298} = $ Gibbs free energy, enthalpy of formation at these conditions

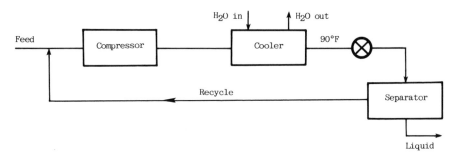

FIGURE T12. Schematic for problem G.

G. Find the pressure needed to obtain maximum liquefaction in the process shown in Figure T12 with the following data on compressibility factors.

Pressure (psia)	Temperature				
	70°F	80°F	90°F	100°F	110°F
700	0.166	0.177	0.415	0.535	0.590
1000	0.223	0.228	0.227	0.253	0.277
1500	0.316	0.319	0.324	0.330	0.338
2000	0.407	0.408	0.410	0.412	0.417
2500	0.494	0.493	0.493	0.495	0.498
3000	0.579	0.577	0.576	0.575	0.577

PROBLEM-SOLVING STRATEGIES

A. To do this problem, I need to know the steps involved in a Rankine cycle with reheating, a plain Rankine cycle, and the Carnot cycle. If I can find the steps defined—possibly in Smith and Van Ness—then I can solve the problem using a set of steam tables. Because of all the parts to this question, I would, in the interests of time, try to find shorter, possibly easier, problems.

B. With the equation for the definition of entropy, part (a) of this problem is simple. Take care with the differing units. For part (b) I must work backwards from the given gaseous entropy at 25°C since I do not know the solid entropy at absolute zero. An easy problem if I am careful with the unit conversions.

C. This equilibrium flash should be simple, because the temperature is specified and the values for K are given. (It would be trial and error if temperature were not specified.) I shall just use component balances

$$FZ_i = LX_i + VY_i \qquad \text{and} \qquad Y_i = K_iX_i$$

so

$$FZ_i = LX_i + VK_iX_i = (L + VK_i)X_i$$

with $L/F = 0.52$, $V/F = 0.48$. I can calculate each X_i in turn.

D. This is a chemical equilibrium problem. The equilibrium constant is given and it can be expressed in terms of total pressure, mole fractions, and fugacity coefficients. To get the mole fractions present at any time, do a material balance with the stoichiometric equation. To get the fugacity coefficients, use the reduced temperatures and pressures of the components with fugacity coefficient graphs. Thus I can solve for conversion if I have expressed mole fractions in terms of conversion. With conversion available, all other calculations are easy. Keep in mind that inerts affect the mole fractions of reactants and products.

E. This problem is a little different. It could be trial and error. I know that at the time that the first drop of liquid forms, the vapor mole fractions are those given. Because of the immiscibility, if water condenses first, its temperature will be at the point that gives a vapor pressure equal to the total pressure times the mole fraction of water vapor, 0.333. If I make this assumption, I must check, using vapor pressures and mole fractions, to see if the hydrocarbon phase could have condensed at a higher temperature. Unless I can find a discussion of this type of problem (Smith and Van Ness or Perry), I would pass over it—with relief.

F. At first glance this looks to be a simple chemical equilibrium problem. All the equations are given to calculate the equilibrium constant. Set up the stoichiometry in terms of conversion and then calculate mole fractions in terms of conversion. Using the equilibrium constant in terms of total pressure, mole fractions, and fugacity coefficients, calculate the conversion. If the equilibrium conversion calculated is greater than 93 %, the claim is possible. Fugacity coefficients can be obtained from a graph in Smith and Van Ness, in terms of reduced temperature and pressures. This should be an easy problem.

G. Liquefaction brings to mind the Joule–Thomson effect, a constant-enthalpy expansion. But to get from enthalpy to compressibility will require using various thermodynamic relationships. I would start with an enthalpy balance around the valve. On the inlet side is all vapor; on the outlet side is vapor and liquid. So enthalpy could be related to the fraction of liquid. I would start this problem, if I felt I had time to think about what thermodynamic relationships are needed.

SOLUTIONS

A. See Figure T13.

Basis: 1-lb circulating steam

Want: Q_{boiler}, Q_{reheat}, Q_{cond}, W_{s1}, W_{s2}, W_p, η, and quality.
Compare to Rankine without reheating and with Carnot.

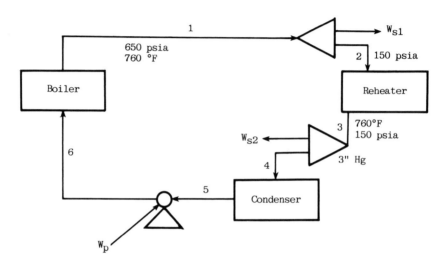

FIGURE T13.   Schematic for solution A(a).

At point (1), $H_1 = 1382.8$, $S_1 = 1.6060$, $T = 760°F$, $P = 650$ psia, $W_{s1} = H_1 - H_2$. For turbine, assume isentropic operation, $S_1 = S_2$. So at point (2), $P_2 = 150$, $S_2 = 1.6060$, $T_2 = 410°F$, $H_2 = 1225$, so

$$W_{s1} = 1383 - 1225 = 158 \text{ Btu/lb}$$

At point (3), $P_3 = 150$ psia, $760°F = T_3$, $S_3 = 1.780$, $H_3 = 1407$,

$$Q_r = H_3 - H_2 = 1407 - 1225 = 182 \text{ Btu/lb}$$

At point (4), $P_4 = 3$ in. Hg, $S_4 = 1.7820$, two-phase region $T_4 = 115°F$.
Let X = fraction liquid

$$S = 1.7820 = X(0.1560) + (1 - X)1.9456 \quad \text{quality} = 1 - X = 90.9\%$$
$$H = ? \quad\quad = X(82.99) + (1 - X)(1111.6) \quad\quad X = 0.091$$

$H_4 = 1018$ Btu/lb; $-W_{s2} = H_4 - H_3 = 1018 - 1407 = -389$ Btu/lb;
$H_5 = 82.99$ Btu/lb.

At point (5), $P_5 = P_4 = 3$ in. Hg $= 1.473$ psia, $T_5 = T_4 = 115°F$, $S_5 = 0.1560$,

$$Q_c = H_5 - H_4 = 83 - 1018 = -935 \text{ Btu/lb}$$

For pump, assume isentropic compression

$$W_p = \int V \, dp \cong V \Delta p = 0.01618(650 - 1.47) = 10.49 \left(\frac{\text{ft}^3}{\text{lb}_m} \right) \left(\frac{\text{lb}_f}{\text{in.}^2} \right)$$

$$= 10.49 \left(\frac{\text{ft}^3}{\text{lb}_m} \right) \left(\frac{\text{lb}_f}{\text{in.}^2} \right) \left(\frac{144 \text{in.}^2}{\text{ft}^2} \right) \left(\frac{1 \text{ Btu}}{778 \text{ ft lb}_f} \right)$$

$$= 1.94 \frac{\text{Btu}}{\text{lb}}$$

At point (6), $H_6 = H_5 + W_p = 83 + 1.94 \cong 85$ Btu/lb

$$Q_b = H_1 - H_6 \cong 1383 - 85 = 1298 \text{ Btu/lb}$$

Check

$$Q_b + Q_r + Q_c = 1298 + 182 - 935 = 545 \text{ Btu/lb}$$

$$W_{s1} + W_{s2} - W_p = 158 + 389 - 2 = 545 \text{ Btu/lb}$$

$$\eta = \frac{\text{work}}{Q_r + Q_b} = \frac{545}{1298 + 182} = \frac{545}{1480} = 37\%$$

Conventional Rankine cycle (see Figure T14).

$$H_A = H_1 = 1382.8 \text{ Btu/lb}, S_A = S_1 = 1.6060, 1.6060 = X(0.1560)$$
$$+ (1 - X)1.9456 \quad \text{or} \quad X = 0.190, \text{quality} = 81\%$$
$$H_B = X(82.99) + (1 - X)1111.6 = 916.1 \text{ Btu/lb}$$
$$W_s = H_A - H_B = 1382.8 - 916.1 = 466.7 \text{ Btu/lb}$$

$$Q_{\text{cond}} = H_c - H_B, \qquad H_c = H_5, \qquad Q_c = 83 - 916.1 = -833.1 \text{ Btu/lb}$$

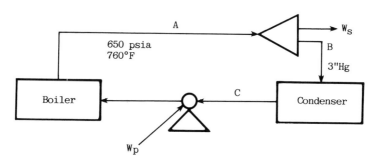

FIGURE T14. Schematic for solution A(b).

As before

$$\int v \, dP = 1.94 \text{ Btu/lb} \quad \text{and} \quad H_D = H_6 = 85 \text{ Btu/lb}$$

$$Q_{\text{boiler}} = H_A - H_D = 1382.8 - 85 = 1297.8 \text{ Btu/lb}$$

$$\eta = \frac{W_{\text{net}}}{Q_{\text{boiler}}} = \frac{466.7 - 1.9}{1297.8} = 35.8\%$$

Carnot efficiency

$$\eta = (760 - 115)/(760 + 460) = 52.9\%$$

B. ΔS for 1 lb NH_3 $\begin{array}{c} 204.4°C \\ 400°F \text{ gas} \end{array} \rightarrow \begin{array}{c} -115°C \\ -175°F \text{ solid} \end{array}$

Gas $400°F \Rightarrow -33.4°C$; $\Delta S = \int (c_p/T) \, dT$;

$$C_p = \alpha + \beta T + \gamma T^2, \quad \alpha = 6.086,$$
$$\beta = 8.813 \times 10^{-3}, \quad \gamma = -1.506 \times 10^{-6}$$

T in K, C_p in Btu/lb mol °F. (Smith and Van Ness, 2nd ed., p. 122)

$$\Delta S = \int_{T_2}^{T_2} \left[\frac{\alpha}{T} + \beta + \gamma T \right] dT = \left(\alpha \ln T + \beta T + \frac{\gamma}{2} T^2 \right)_{T_1}^{T_2}$$

$$= \alpha \ln \frac{T_2}{T_1} + \beta(T_2 - T_1) + \frac{\gamma}{2}(T_2^2 - T_1^2)$$

$T_1 = 400°F = 477 \text{ K}, \, T_2 = -33.4°C = 239.6 \text{ K, so}$

$$\Delta S_1 = -6.1645 \frac{\text{Btu}}{\text{lb mol °R}} \div \frac{17 \text{ lb}}{\text{lb mol}} = -0.3626 \frac{\text{Btu}}{\text{lb °R}}$$

Gas → liquid at $-33.4°C$

$$\Delta S_2 = \frac{-\Delta H_{\text{vap}}}{T_{\text{boil}}} = \frac{-5581 \text{ cal}}{\text{g mol } 239.6 \text{ K}} = -23.29 \frac{\text{cal}}{\text{g mol K}}$$

$$= -23.29 \frac{\text{Btu}}{\text{lb mol °R}} \div \frac{17 \text{ lb}}{\text{lb mol}} = -1.370 \frac{\text{Btu}}{\text{lb°F}}$$

Liquid $-33.9°C \rightarrow -77.7°C$

$$\Delta S_3 = \int_{239.6}^{195.3} \frac{C_p}{T} \, dT; \quad C_p = 1.06 \frac{\text{cal}}{\text{g K}}$$

$$= C_p \ln \frac{195.3}{239.6} = -0.217 \frac{\text{cal}}{\text{g K}} = -0.217 \frac{\text{Btu}}{\text{lb°F}}$$

Liquid → solid at $-77.7°C$

$$\Delta S_4 = \frac{-\Delta H_{fus}}{T_{melt}} = \frac{-1352 \text{ cal}}{\text{g mol } 195.3 \text{ K}} = -6.923 \frac{\text{cal}}{\text{g mol}°C}$$

$$= -6.923 \frac{\text{Btu}}{\text{lb mol}°F} \bigg/ 17 \frac{\text{lb}}{\text{mol}} = -0.407 \frac{\text{Btu}}{\text{lb}°F}$$

Solid $-77.7°C \to -115°C$

$$\Delta S_5 = \int_{195.3}^{158} \frac{C_p}{T} dT, \qquad C_p = 0.502 \frac{\text{cal}}{\text{g K}}$$

$$= 0.502 \ln \frac{158}{195.3} = -0.1064 \frac{\text{Btu}}{\text{lb}°F}$$

$$\Delta S = -0.3626 - 1.370 - 0.217 - 0.407 - 0.1064$$

$$= -2.463 \frac{\text{Btu}}{\text{lb}°F} \qquad \text{answer to (a)}$$

Absolute entropy of solid NH_3 at melting point $-77.7°C$. Note that we cannot use $\int_0^{T_{melt}}(C_p/T) \, dT$ with $C_p = 0.502$ cal/g K, because we get an indeterminate term at $T = 0$. So we use the summation scheme shown in Figure T15.
So $S - \Delta S_4 - \Delta S_3 - \Delta S_2 + \Delta S_n = \Delta S$ or $S = \Delta S - \Delta S_n + \Delta S_2 + \Delta S_3 + \Delta S_4$

$$\Delta S_n = \alpha \ln \frac{298}{239.6} + \beta(298 - 239.6) + \frac{\gamma}{2}(298^2 - 239.6^2)$$

$$= 1.8185 \frac{\text{Btu}}{\text{lb mol}°F} = \frac{1.8185}{17} = 0.1070 \frac{\text{cal}}{\text{g K}} \quad \text{or} \quad \frac{\text{Btu}}{\text{lb}°R}$$

$$S = \frac{46.03}{17} \frac{\text{Btu}}{\text{lb}°F} - 0.107 - 1.370 - 0.217 - 0.407$$

$$= 0.6066 \frac{\text{Btu}}{\text{lb}°F} \qquad \text{answer (b)}$$

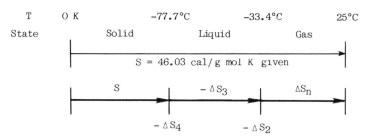

FIGURE T15. Schematic for solution B.

C. *Basis:* 1-mol feed; $y_i V + x_i L = Z_i$ with $y_i = K_i x_i$, so

$$K_i x_i V + x_i L = Z_i \qquad x_i = \frac{Z_i}{K_i V + L}$$

and

$$L = 0.52 \qquad V = 0.48$$

Compound	Z_i	K_i	$0.48\,K_i + 0.52$	x_i	y_i
Methane	0.15	17.0	8.68	0.0173	0.2938
Ethane	0.15	3.1	2.008	0.0747	0.2316
Propane	0.25	1.0	1.00	0.25	0.25
Isobutane	0.20	0.44	0.7312	0.2735	0.1204
n-Butane	0.20	0.32	0.6736	0.2969	0.0950
n-Pentane	0.05	0.096	0.5661	0.0883	0.0085
				1.0007	0.9992

D. *Basis:* 1-mol entering gas or 0.3 mol CO, 0.7 mol N_2.

$$\text{FeO(s)} + \text{CO(g)} \rightleftharpoons \text{Fe(s)} + \text{CO}_2\text{(g)} \qquad T = 100°C; P = 100 \text{ atm.}$$

At $t = 0$, CO $= 0.3$; $CO_2 = 0$.
At $t = t$, CO $= C_{CO}$; $CO_2 = C_{CO_2}$; $0.3 - C_{CO} = C_{CO_2} = x$, then $C_0 = 0.3 - \cup\bigcup x$; total moles $= C_{CO} + C_{CO_2} + 0.7 = 0.3 - x + x + 0.7 = 1$; mole fraction CO $= 0.3 - x$, mf $CO_2 = x$,

$$K_a = \frac{a_{Fe} a_{CO_2}}{a_{FeO} a_{CO}} = \frac{a_{CO_2}}{a_{CO}} = \frac{y_{CO_2}}{y_{CO}} \frac{\phi_{CO_2}}{\phi_{CO}} = 35$$

Get ϕs from Smith and Van Ness, p. 354, 2nd ed.

	$T_C(K)$	$P_c(\text{atm})$	T_r	P_r	ϕ
CO_2	304.1	73	1.23	1.37	0.80
CO	134	35	2.78	2.86	0.97

So

$$\frac{y_{CO_2}}{y_{CO}} = 35 \frac{0.97}{0.80} = 42.44$$

Then

$$42.44 = (x)/(0.3 - x) \qquad \text{or} \qquad x = 0.293 \text{ (also moles of Fe produced)}$$

For pure CO entering

$$\text{mf } CO = 1 - x, \qquad \text{mf } CO_2 = x$$

$$\frac{x}{1-x} = 42.44 \qquad \text{or} \qquad x = 0.977 \text{ (also moles Fe produced)}$$

$$\text{moles gas}\left(\frac{25°C}{1 \text{ atm}}\right)\bigg/1000 \text{ ft}^3 = (1000 \text{ ft}^3)\bigg/\left[359 \frac{\text{ft}^3}{\text{mol}}\left(\frac{298}{273}\right)\right] = 2.55$$

For 30% CO: Fe produced = 0.293(2.55)(55.85) = 41.73 lb.
For 100% CO: Fe produced = 0.977(2.55)(55.85) = 139.14 lb.

E. Composition: $0.583C_6, 0.083C_7, 0.333H_2O$; H_2O and hydrocarbons are immiscible as liquids; cooling down from 350°F at moment of first drop of liquid forming $y_6 = 0.583$, $y_7 = 0.083$, $y_w = 0.333$. So

$$\begin{array}{lll}
P_6 = 0.583 \text{ atm} & P_7 = 0.083 \text{ atm} & P_w = 0.333 \\
\quad = 443.33 \text{ mm Hg} & \quad = 63.33 \text{ mm Hg} & \quad = 253.33 \text{ mm Hg} \\
& & \quad = 4.90 \text{ psia}
\end{array}$$

If H_2O condenses first, T will be that at which $VP_w = 4.90$ psia or $T = 161°F = 621°R$. Check for hc condensation:

$$y_6 P_T = x_6 VP_6; \; y_7 P_T = x_7 VP_7$$

$$x_6 = y_6\left(\frac{P_T}{VP_6}\right); \; x_7 = y_7\left(\frac{P_T}{VP_7}\right); \; x_6 + x_7 = 1$$

$$= \frac{443.33}{VP_6}; \qquad = \frac{63.33}{VP_7}$$

$$\ln VP_6 = 17.7109 - \frac{6816.4}{T}, \qquad \ln VP_7 = 17.9184 - \frac{7547.4}{T}$$

At 621°R

$$VP_6 = 840.8, \qquad x_6 = 0.5272$$

At 621°R

$$VP_7 = 318.9, \qquad x_7 = 0.1986, \qquad x_6 + x_7 = 0.7258$$

So there is no condensation of hydrocarbons. First condensation occurs at 161°F (answer 1) and it is pure H_2O (answer 2).

Assume $T = 150°F$ (610°R) for condensation of 2nd phase. At this T,

$$VP_w = 3.718 \text{ psia (192.22 mm Hg)},$$

$$\bar{P}_{hc} = 760 - 192.22 = 567.78 \text{ mm Hg},$$

$$y_{hc} = \frac{567.78}{760} = 0.747.$$

Thus

$$\frac{y_6}{y_7} = \frac{0.583}{0.083} = 7 \quad \text{and} \quad y_6 + y_7 = 0.747$$

So

$$y_6 = \frac{7}{8}(0.747) = 0.6536, \quad y_7 = \frac{0.747}{8} = 0.0934$$

$$VP_6 = 689.85, \qquad\qquad VP_7 = 256.11$$

$$x_6 = 0.7201, \qquad\qquad x_7 = 0.2772 \qquad \sum = 0.9972$$

Assume $T = 149°F(609°R)$. At this T, $VP_w = 3.627$ psia (187.52 mm Hg),

$$\bar{P}_{hc} = 760 - 187.52 = 572.48, \; y_{hc} = 572.78/760 = 0.753.$$

So

$$y_6 = \frac{7}{8}(0.753) = 0.6591, \quad y_7 = \frac{0.753}{8} = 0.0942$$

$$VP_6 = 677.31, \qquad\qquad VP_7 = 250.96$$

$$x_6 = 0.7396, \qquad\qquad x_7 = 0.2853 \qquad \sum = 1.0248$$

So to the nearest degree, the second phase condenses at 150°F (answer 3) and its composition is 72.01 % C_6 and 27.72 % C_7 (answer 4).

F. $C_2H_6(g) \rightleftarrows C_2H_4(g) + H_2(g)$ $\qquad$ 1 $\qquad$ 0 $\qquad$ 0 $\qquad$ Conc. at $t = 0$;

$$\qquad\qquad\qquad\qquad\qquad\qquad 1-x \quad x \quad x \qquad\qquad t = t:$$

$$\text{Total moles} = 1 + x$$

$$\Delta G_{298}^\circ = (\Delta G_{298\,H_4}^\circ + \Delta G_{298\,H_2}^\circ)_f - (\Delta G_{298\,H_6}^\circ)$$
$$= 16{,}282 + 0 - (-7860) = +24{,}142 \text{ cal/mol}$$
$$\Delta H_{298}^\circ = (\Delta H_{298\,H_4}^\circ + \Delta H_{298\,H_2}^\circ)_f - (\Delta H_{298\,H_6}^\circ)_f$$
$$= 12{,}500 + 0 - (-20{,}236) = +32{,}736 \text{ cal/mol}$$

$$\Delta H_T^\circ = \Delta H_{298}^\circ + \int_{298}^{T} \left(\sum v_i C_{Pi} \right) dT$$

$$= 32{,}736 + \int_{298}^{T} (2.8 + 0.03\,T + 6.9 + 0.004\,T - 2.3 - 0.02\,T)\, dT$$

$$= 32{,}736 + \int_{298}^{T} (7.4 + 0.014\,T)\, dT$$

$$= 32{,}736 + 7.4(T - 298) + 0.007(T^2 - 298^2)$$

Now $\Delta G^\circ = -RT \ln K_a$, so at $T = 298$ K, $\ln K_a = -24{,}142/1.987(298) = -40.77$. Now

$$\frac{d \ln K_a}{d(T)} = + \frac{\Delta H^\circ}{RT^2} \quad \text{(van't Hoff)}$$

$$\Delta H_T^\circ = 29{,}909.17 + 7.4\,T + 0.007\,T^2$$

$$\Delta H_T^\circ/T^2 = 29{,}909.17\,T^{-2} + (7.4/T) + 0.007$$

$$R \int_{\ln K_{a298}}^{\ln K_{aT}} d \ln K_a = \int_{298}^{T} \left(29{,}909.17\,T^{-2} + \frac{7.4}{T} + 0.007\right) dT$$

$$R(\ln K_{aT} - \ln K_{a298}) = \left(\frac{-29{,}909.17}{T} + 7.4 \ln T + 0.007\,T\right)_{298}^{T}$$

$$R(\ln K_{aT} + 40.77) = \left[-29{,}909.17\left(\frac{1}{T} - \frac{1}{298}\right) + 7.4 \ln \frac{T}{298}\right.$$
$$\left. + 0.007(T - 298)\right]$$

$T = 1000$ K

$$R(\ln K_{aT} + 40.77) = 84.33, \qquad \ln K_{aT} + 40.77 = 42.44$$

$$\ln K_{aT} = 1.67, \qquad K_{a\,1000} = 5.317 = \frac{y_{H_2} y_{H_4} P^2}{y_{H_6} P}$$

or

$$\frac{y_{H_2} y_{H_4}}{y_{H_6}} = 53.17, \qquad mf_{H_2} = \frac{x}{1-x}, \qquad mf_{H_4} = \frac{x}{1-x}, \qquad mf_{H_6} = \frac{1-x}{1+x}$$

$$53.17 = \frac{x^2(1+x)}{(1+x)^2(1-x)} = \frac{x^2}{1-x^2}, \qquad X^2 = 0.9815,$$

$$x = 0.9907 \text{ equil conv.}$$

So a conversion of 93 % is possible.

G. See Figure T16. This is a Joule–Thomson liquefaction problem.
Basis: 1-lb gas at 90°F and $P = ?$ entering cracked valve. $H_{in} = H_{out}$;
$H_{in} = x H_{sat\,liq} + (1 - x) H_{sat\,vap}$, or

$$x = \frac{H_{in} - H_{sat\,vap}}{H_{sat\,liq} - H_{sat\,vap}}$$

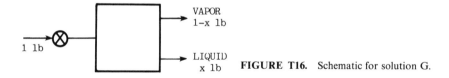

FIGURE T16. Schematic for solution G.

We want

$$\left(\frac{\partial x)}{\partial P}\right)_{90°F} = \left(\frac{\partial x}{\partial P}\right)_T = 0 = \left(\frac{\partial H_{in}}{\partial P}\right)_{T=90°F}\left(\frac{1}{H_{sat\ liq} - H_{sat\ vap}}\right)$$

so we want

$$\left(\frac{\partial H}{\partial P}\right)_{T=90°F} = 0$$

Now $dH = T\,dS + V\,dP$. So

$$\left(\frac{\partial H}{\partial P}\right)_{T=90°} = V + T\left(\frac{\partial S}{\partial P}\right)_T$$

from Maxwell relations

$$-\left(\frac{\partial S}{\partial P}\right)_T = \left(\frac{\partial V}{\partial T}\right)_P$$

$$\left(\frac{\partial H}{\partial P}\right)_{T=90°} = V - T\left(\frac{\partial V}{\partial T}\right)_P$$

But

$$V = (zRT)/P$$

So

$$\left(\frac{\partial V}{\partial T}\right)_P = \frac{R}{P}\left[\frac{\partial(ZT)}{\partial T}\right]_P = \frac{R}{P}\left[Z + T\left(\frac{\partial Z}{\partial T}\right)_P\right]$$

So

$$\left(\frac{\partial H}{\partial P}\right)_{T=90°} = V - \frac{RT}{P}\left[Z + T\left(\frac{\partial Z}{\partial T}\right)_P\right] = \frac{ZRT}{P} - \frac{ZRT}{P} - \frac{RT^2}{P}\left(\frac{\partial Z}{\partial T}\right)_P$$

Then

$$-\frac{RT^2}{P}\left(\frac{\partial Z}{\partial T}\right)_P = 0 \quad \text{or} \quad \left(\frac{\partial Z}{\partial T}\right)_P = 0$$

From data

$$\text{at } P = 3000 \text{ psia}, \quad \partial Z/\partial T = 0 \text{ at } T = 100°F$$
$$\text{at } P = 2500 \text{ psia}, \quad \partial Z/\partial T = 0 \text{ at } T = \ \ 85°F$$
$$\text{at } P = 2000 \text{ psia}, \quad \partial Z/\partial T = 0 \text{ at } T = \ \ 70°F$$

So at $T = 90°F$ use $\frac{5}{15}$ of $(3000 - 2500)$ added to 2500 or $P = 2667$ psia

PROCESS DESIGN

I. INTRODUCTION

Most of the process design problems found on the test are really just unit operation problems covered under other sections of this book. Therefore this chapter will cover areas of process design not addressed elsewhere. These topics include stoichiometry, heat and material balances, optimization, and control.

II. STOICHIOMETRY

Stoichiometry problems involve chemical balances. Combustion problems are based on stoichiometry principles—the relationships of gas volume, mass, and composition in a chemical reaction. So we use combustion problems to get you involved, once again, with stoichiometry. Figure D1 represents most combustion problems.

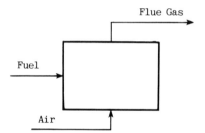

FIGURE D1. Combustion problem schematic.

All combustions use oxygen, but costs usually make pure oxygen unattractive. Thus air is used. For calculation purposes, air is a mixture of 79 volume % N_2 and 21 volume % O_2, giving a nitrogen to oxygen ratio of 79/21 or 3.76.

The flue gas is most often analyzed by the Orsat technique. This is a simple volumetric method, but it does not account for water or nitrogen oxides. Thus, an Orsat analysis is given on a moisture free basis. Because the Orsat cannot account for nitrogen oxides, we assume that nitrogen oxides are negligible. For the analysis, we say that none of the nitrogen in the air reacts. Therefore, for calculation purposes, all the nitrogen entering the system in the air stream leaves as part of the flue gas.

To insure complete combustion of all the carbon and available hydrogen in the fuel, air is always fed in excess of the amount needed. So with complete combustion, all carbon in the fuel shows up in the flue gas as either carbon monoxide or carbon dioxide. The theoretical oxygen is that needed to burn all carbon in the fuel to *carbon dioxide*. We calculate the percentage of excess air from

$$\text{percentage excess air} = \frac{O_2 \text{ supplied} - O_2 \text{ theoretical}}{O_2 \text{ theoretical}} \, 100$$

Example Problem:

Fuel	Air (vol)	Flue gas	(Moisture Free Basis)(mol)
Carbon	79% N_2	CO_2	X
Hydrogen	21% O_2	CO	Y
Moisture		O_2	Z
Ash		N_2	$100 - X - Y - Z$
Other			

Combustion equation:

$$C_A H_B + M O_2 + P N_2 \longrightarrow X CO_2 + Y CO + Z O_2$$
$$+ (100 - X - Y - Z)N_2 + R H_2O$$

Relationships from the combustion equation:

$$
\begin{aligned}
\text{Carbon} \quad & A = X + Y \\
\text{Hydrogen} \quad & B = 2R \\
\text{Oxygen} \quad & M = X + \tfrac{1}{2}Y + Z + \tfrac{1}{2}R \\
\text{Nitrogen} \quad & P = 100 - X - Y - Z
\end{aligned}
$$

Calculation: Always choose and write down the basis
Basis: 100 mol of dry flue gas

1. Air feed rate.

$$N_2 \text{ in flue gas} = 100 - X - Y - Z = P$$

$$\frac{\text{air feed}}{100 \text{ mol flue gas}} = \frac{100 - X - Y - Z}{0.79} = \frac{P}{0.79}$$

$$O_2 \text{ in air feed} = 0.21\left(\frac{P}{0.79}\right) = \frac{P}{3.76}$$

2. Calculate composition of feed.
    (a) Carbon in flue gas = carbon in fuel.

$$\text{moles C} = A = X + Y$$

$$C = CO_2 + CO$$

$$\text{pounds of C} = (X + Y)\ 12\left(\frac{\text{lb}}{\text{lb mol}} \atop \text{MW}\right)$$

    (b) Hydrogen in fuel = hydrogen in water.

$$\text{moles } H_2 = \tfrac{1}{2}B = R$$

(c) Oxygen balance gives the moles of water. Moles O_2 in air = moles O_2 in flue gas + moles O_2 used to form CO_2, CO, and H_2O. So

$$M = 0.21\left(\frac{100 - Z - Y - Z}{0.79}\right) = X + \tfrac{1}{2}Y + Z + \tfrac{1}{2}R$$

or

$$\tfrac{1}{2}R = (0.21/0.79)(100 - X - Y - Z) - (X + \tfrac{1}{2}Y + Z)$$

$$R = 2[(0.21/0.79)(100 - X - Y - Z) - (X + \tfrac{1}{2}Y + Z)]$$

$$\text{pounds } H_2 \text{ in feed} = R\left(\frac{2\,\dfrac{\text{lb}}{\text{lb mol}}}{\text{MW}}\right)$$

Percentage excess air:

O_2 supplied $= (0.21/0.79)(100 - X - Y - Z)$

O_2 theoretical $= X + Y + \tfrac{1}{2}R$

Note that we do not use $\tfrac{1}{2}Y$ because theoretical oxygen is based upon complete combustion.

$$\text{percentage excess} = \frac{\text{supplied} - \text{theoretical}}{\text{theoretical}}\,100$$

Some Words on Gas Volumes. At standard conditions (STP) of 0°C and 1 atm, 1 lb mol of dry gas occupies 359 ft^3. For other than standard conditions we use the ideal gas law $PV = nRT$

where P = absolute pressure,
      V = volume,
      n = number of moles,
      R = gas law constant, and
      T = absolute temperature.

So

$$\frac{PV}{P_{\text{std}}V_{\text{std}}} = \frac{nRT}{nRT_{\text{std}}} \quad \text{or} \quad V = V_{\text{std}}\left(\frac{P_{\text{std}}}{P}\right)\left(\frac{T}{T_{\text{std}}}\right)$$

III. MATERIAL AND ENERGY BALANCES

A. Introduction

All balances, whether material or energy, start with the fundamental balance equation,

$$\text{input} - \text{output} + \text{production} = \text{accumulation}$$

Since unsteady-state problems on the test are rare, we shall assume steady state, or

$$\text{input} - \text{output} + \text{production} = 0$$

The production term is of importance only for problems involving chemical reaction. Without reactions, both material and energy balances are determined from

$$\text{input} - \text{output} = 0$$

B. A Typical Problem

Consider the process, shown in Figure D2, in which streams A and B react to form stream C. If the temperatures of streams A and B are given, what is the temperature of stream C?

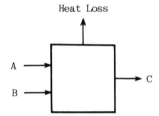

FIGURE D2.   Typical process energy balance.

We must first do a material balance, using the reaction stoichiometry. This will determine the amounts of each of the components in each of the streams. To do an energy balance, we need the enthalpy of each stream. Actually, the enthalpy of component x is given by

$$H_x = X \int_{T_0}^{T} C_p \, dT$$

where H_x = enthalpy (Btu),
 X = amount of x (lb),
 C_p = specific heat of x (Btu/lb°F),
 T = temperature of x (°F), and
 T_0 = base temperature (chosen) °F.

Practically, most design work uses a mean C_p, which is defined by

$$C_{p\,\text{mean}}(T - T_0) = \int_{T_0}^{T} C_p \, dT$$

Plots of mean C_p versus temperature T can be found in most thermodynamics texts. So the enthalpy of stream A is

$$H_A = \left(\sum W_A C_{p\,\text{mean}}\right)(T_A - T_0)$$

Likewise

$$H_B = \left(\sum W_B C_{p\,mean}\right)(T_B - T_0)$$
$$H_C = \left(\sum W_C C_{p\,mean}\right)(T_C - T_0)$$

where the W_i are the weight fractions of the components in the stream.

If we know the heat loss H_L and the heat of reaction ΔH_R at temperature T, then the energy balance becomes[†]

$$H_A + H_B - H_C - H_L - \Delta H_R = 0$$

This equation is solved for the temperature T.

C. Heats of Formation and Combustion

The heat of formation ΔH_f is the heat used $(+)$ or evolved $(-)$ in forming one mole of a compound from the elements that make it up. The heat of combustion ΔH_c is the heat used $(+)$ or evolved $(-)$ in burning with oxygen one mole of a compound to form specified combustion products. Standard tables of these data, usually at 25°C, are available.

For example,

$$\tfrac{1}{2}N_2(g) + \tfrac{1}{2}O_2(g) \longrightarrow NO(g), \qquad \Delta H_f = +21.6 \text{ kcal/g mol}$$
$$\tfrac{1}{2}N_2(g) + O_2(g) \longrightarrow NO_2(g), \qquad \Delta H_f = +8.04 \text{ kcal/g mol}$$

The first equation indicates that if and when 1 g mol of nitric oxide (gas) at 1 atm and 25°C is formed from $\tfrac{1}{2}$ g mol each of nitrogen (gas) and oxygen (gas), both at 25°C and 1 atm, 21.6 kcal must be added to the reaction in order to form the nitric oxide (or 21.6 kcal must be consumed by the reaction). When the equations are written as above, the $+$ sign means that heat is used in the reaction (called an endothermic reaction, that is, heat is put "endo" the reaction) and a $-$ sign means that heat is given off by the reaction (called an exothermic reaction, that is, the heat "exits" the reaction).

An easy way of working with heats of reaction is to write the two preceding equations in the following way:

$$\tfrac{1}{2}N_2(g) + \tfrac{1}{2}O_2(g) + 21.6 \text{ kcal/g mol NO} \longrightarrow NO(g)$$
$$\tfrac{1}{2}N_2(g) + O_2(g) + 8.04 \text{ kcal/g mol NO}_2 \longrightarrow NO_2(g)$$

Since most heats of formation (and heats of combustion) are tabulated at 25°C and 1 atm pressure, we would seem to be in trouble. In fact, it is doubtful if the preceding reactions would even work at 25°C and 1 atm. We shall resolve this shortly.

[†] Note that the production term, the heat of reaction, has a negative sign because the convention is that an exothermic reaction (one losing heat to the surroundings) carries a negative sign. So for a heat-producing reaction, the term $-\Delta H_R$ is a positive amount of heat.

D. Heat of Reaction

1. At 25°C

The standard heat of reaction, ΔH_R, is the difference between the enthalpy of the products and the enthalpy of the reactants when the reaction both begins and ends at 25°C and 1 atm. If the standard heat of reaction cannot be found, it can be calculated from standard heats of formation and/or standard heats of combustion (if you can find them).

To illustrate, what is the standard heat of reaction for the reaction:

$$NO(g) + \tfrac{1}{2}O_2(g) \longrightarrow NO_2(g)?$$

Using the previous two equations

$$\tfrac{1}{2}N_2(g) + O_2(g) + \; 8.04 \, \text{kcal} \longrightarrow NO_2(g)$$
$$\tfrac{1}{2}N_2(g) + \tfrac{1}{2}O_2(g) + 21.6 \, \text{kcal} \longrightarrow NO(g)$$

and subtracting the second from the first, using the $\rightarrow$ as an equal sign, and using algebraic principles:

$$\tfrac{1}{2}O_2(g) - 13.56 \, \text{kcal} \rightarrow NO_2(g) - NO(g)$$

or

$$NO(g) + \tfrac{1}{2}O_2(g) - 13.56 \, \text{kcal} \rightarrow NO_2(g)$$

So the standard heat of reaction is

$$\Delta H_R = -13.56 \, \text{kcal/g mol NO}$$

or 13.56 kcal are evolved for each gram mole of NO_2 formed.

For those who prefer to memorize formulas,

$$\Delta H_R = \Delta H_f(\text{products}) - \Delta H_f(\text{reactants})$$

or

$$\Delta H_R = \Delta H_c(\text{reactants}) - \Delta H_c(\text{products})$$

2. At Other Temperatures

The standard heats are tabulated at 1 atm and 25°C (sometimes 18°C). We shall need the heat of reaction at other conditions. Now most thermodynamic functions are point functions; they are functions of the final and initial states only. So, in Figure D3, the enthalpy change in going from point 1 directly to point 4 is equivalent to that going from 1 to 2 to 3 to 4.

Equating the enthalpy balances for these two schemes gives

$$\Delta H_{r1} = (H_2 - H_1)_{\text{reactants}} + \Delta H_{r2} + (H_1 - H_2)_{\text{products}}$$

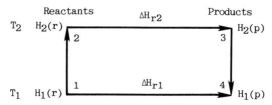

FIGURE D3. Heat of reaction at different temperatures.

or

$$\Delta H_{r2} = \Delta H_{r1} - (H_2 - H_1)_{\text{reactants}} - (H_1 - H_2)_{\text{products}}$$

or

$$\Delta H_{r2} = \Delta H_{r1} - (H_2 - H_1)_{\text{reactants}} + (H_2 - H_1)_{\text{products}}$$

IV.   OPTIMIZATION

A.   Introduction

All chemical engineering designs are optimized, whether formally or informally. The main difficulty is determining the objective function—the quantity that is to be minimized or maximized. Fortunately, for testing purposes this is given. Most practical problems have several variables that affect the optimum. In the interest of practical test taking, single-variable optimization is reasonable. Most of these problems can be solved either graphically or analytically.

B.   Graphical Method

A graphical solution is straightforward if the objective function, such as total cost, can be calculated as a function of an independent variable, such as insulation thickness, which can be varied to affect the total cost. Consider the cost of insulating a pipe. The more insulation you use the higher the cost. But the more insulation you use the less heat is lost from the pipe. For example, if the cost of lagging as a function of thickness X is given by

$$\$_C = X$$

and the cost of the heat loss is given by

$$\$_L = 1 + 4/X$$

we can plot both curves and add them together for the total cost. We see from Figure D4 that the minimum total cost is \$5 at an insulation thickness of 2.

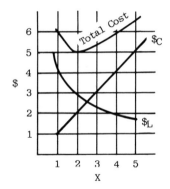

FIGURE D4.   Cost of insulating a pipe.

C. Analytical Method

1. General Conditions

When we plot our optimization problem, we see how the objective function varies as the independent variable changes. We can see if anything unusual is occurring. When we solve the problem analytically, we must consider whether the objective function has any discontinuities or if the independent variable is bounded.

Solving optimization problems analytically we must be aware that the optimum, if one exists, can occur only at:

•   a boundary of the independent variable,
•   a discontinuity in the objective function, or
•   a stationary point of the objective function.

All possibilities must be evaluated for each problem. Of these possibilities, only the methods for finding stationary points will be discussed.

2. Unconstrained Optimization

For continuous functions such as those shown in Figure D5, the slopes, or derivatives, are also continuous. For curve a, to the left of the maximum the derivative is positive; to the right of the maximum the derivative is negative. At the

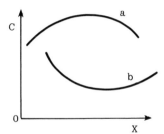

FIGURE D5.   Optima.

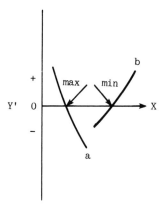

FIGURE D6. Derivatives of optima.

maximum point, the derivative is zero. The opposite is true for curve b with the minimum.

So to find the stationary points analytically, we set the first derivative of the objective function to zero. The resulting independent-variable values are maxima, minima, or inflection points. If we plot the first derivatives of curves a and b (Figure D6), we find that the curve with the maximum has a negative second derivative at the stationary point and the minimum curve has a positive second derivative at the stationary point. So analytically, we evaluate the second derivatives at the stationary points.

The total cost of insulation was

$$C = 1 + \frac{4}{X} + X$$

The first derivative with respect to X is

$$C' = -\frac{4}{X^2} + 1$$

Setting $C' = 0$ and solving for X

$$X^2 = 4$$

or

$$X = \pm 2$$

Since we cannot have negative thicknesses of insulation,

$$X = 2$$

is a stationary point. Taking the second derivative of the total cost gives

$$C'' = 8/X^3$$

The second derivative, evaluated at the stationary point, is

$$C''(2) = 1$$

a positive number. Thus the stationary point, $X = 2$, gives the *minimum* total cost.

3. Constrained Optimization

It often happens in engineering problems that an objective function is to be minimized,

$$\text{minimize } C(x, y)$$

but at the same time, the independent variables are constrained by an equation such as

$$f(x, y) = 0$$

There are several methods of solving this type of problem.

a. Substitution

If easily done, it is always best to solve the constraint equation for one variable in terms of the others. For our case, we might be able to get

$$y = g(x)$$

easily from

$$f(x, y) = 0$$

We then substitute this equation for y into the function to be minimized,

$$\text{minimize } C\{x, g(x)\}$$

Now the cost function depends only upon x and we find the minimum as before.

b. Lagrangian Multiplier

For situations in which it is difficult to solve the constraint equation for one variable in terms of the others, we use the Lagrangian multiplier method. With this method we add to the original objective function, the constraint equation multiplied by a constant λ

$$\bar{C} = C(x, y) + \lambda f(x, y)$$

To find the optimum, we calculate the first derivative of $\bar{C}$ with respect to each independent variable and set it to zero, or

$$\frac{\partial \bar{C}}{\partial x} = \frac{\partial C}{\partial x} + \lambda \frac{\partial f}{\partial x} = 0$$

$$\frac{\partial \bar{C}}{\partial y} = \frac{\partial C}{\partial y} + \lambda \frac{\partial f}{\partial y} = 0$$

This gives two equations for the three unknowns x, y, and λ. The third equation is

$$f(x, y) = 0$$

There is no easy method of telling whether the solution is a maximum, minimum, or inflection point. The easiest method is to evaluate the objective function at locations near the solution to confirm that the solution is indeed a minimum (if you were hoping to minimize an objective function).

V. PROCESS CONTROL

A. Introduction

Processes are designed to operate at steady-state conditions. They never do. All processes change with time for many reasons: the feed rate, composition, or temperature is usually variable; the steam temperature or rate varies; or the atmospheric conditions change.

To introduce process control, we consider the well-stirred vessel in Figure D7 of volume V into which is flowing a fluid of concentration $x(t)$ at a rate of q. Leaving the vessel at rate q is the fluid of concentration $y(t)$. An unsteady-state material balance gives

$$V\frac{dy}{dt} = qx - qy$$

If the input concentration suddenly changes from $x = 0$ to $x = a$ (a step function), this first-order differential equation can be solved to give

$$y(t) = a[1 - e^{-qt/V}]$$

Note that the two constants that define the system, V and q, appear only as a ratio. This ratio is called the time constant τ since it has units of time

$$\tau = V/q$$

With the time constant notation, the concentration equation is

$$y(t) = a[1 - e^{-t/\tau}]$$

B. The Laplace Transform

Most chemical engineering control problems involve keeping a process operating near a steady-state point. If the control scheme is a good one, large devia-

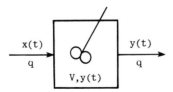

FIGURE D7. A well-stirred vessel.

tions from the operating point are unlikely. Small deviations from the setpoint allow one to assume that most control problems are linear systems. The time-honored method for solving linear control problems is the Laplace transformation.

The Laplace transform of the function $f(t)$ is defined by

$$L[f(t)] = \int_0^\infty f(t)e^{-st}\, dt$$

So we convert a function of time t to a function of s, or

$$L[f(t)] = F(s)$$

Taking the Laplace transform of

$$f(t) = e^{-at}$$

we obtain, through a formal integration,

$$L[f(t)] = L[e^{-at}] = \frac{1}{s+a}$$

Note that when the denominator is zero,

$$s = -a$$

the coefficient of t in the exponential term. Using integration by parts, we find that

$$L\left[\tau\left(\frac{dy}{dt}\right)\right] = \tau[sY(s) - y(0)]$$

where $Y(s)$ is the Laplace transform of $y(t)$. Many texts contain the pairs of Laplace transforms and operations.

C.   Transfer Functions

Returning to our original stirred tank, by defining

$$L[y(t)] = Y(s), \qquad L[x(t)] = X(s)$$

our differential equation in time becomes an algebraic equation in s,

$$\tau s Y(s) + Y(s) = X(s)$$

or

$$Y(s) = \left(\frac{1}{\tau s + 1}\right)X(s)$$

or, in general,

$$Y(s) = G(s)X(s)$$

The bracketed term that defines $G(s)$ is called the transfer function of the system.

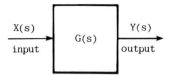

FIGURE D8.   General block diagram.

For any given input $x(t)$ to our stirred tank, we could look up its transform $X(s)$. The transform of the output $Y(s)$ is then known; the time response can be found from Laplace transform tables. For a step input in $x(t)$

$$x(t < 0) = 0, \qquad x(t > 0) = a$$

the transform is

$$X(s) = a/s$$

Thus,

$$Y(s) = \left(\frac{1}{\tau s + 1}\right)\frac{a}{s} = \frac{a/\tau}{s(s + 1/\tau)}$$

$$= \frac{a}{s} - \frac{a}{s + 1/\tau}$$

The inverse transform of $Y(s)$,

$$L^{-1}[Y(s)] = y(t)$$

is

$$y(t) = a[1 - e^{-t/\tau}]$$

D.   Block Diagrams

One of the advantages of the transfer function is that it leads to block diagrams. The stirred tank, in block diagram form, is

$$Y(s) = G(s)X(s)$$

as shown in Figure D8, or

$$Y(s) = \left(\frac{1}{\tau s + 1}\right)X(s)$$

as shown in Figure D9.

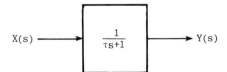

FIGURE D9.   Stirred tank block diagram.

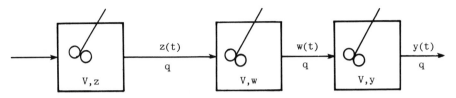

FIGURE D10. Three stirred tanks in series.

We can extend this concept. Consider the three stirred tanks arranged in series in Figure D10.

This cascade is represented in block-diagram notation by Figure D11 or by

$$Z(s) = G_1(s)X(s)$$
$$W(s) = G_2(s)Z(s)$$
$$Y(s) = G_3(s)W(s)$$

which, when combined algebraically, gives

$$Y(s) = G_3W = G_3G_2Z = G_3G_2G_1X(s)$$

E. Closing the Loop

Consider the closed-loop (feedback) control system shown in Figure D12 where the Laplace-transformed variables are

$R(s) =$ set point,
$E(s) =$ error between set point and Y,
$G_c =$ controller transfer function,
$P(s) =$ output from controller (usually pressure),
$G_v =$ valve transfer function,
$X(s) =$ output of valve, same units as Y,
$U(s) =$ outside disturbance to process,
$G =$ process transfer function,
$Y(s) =$ process output variable,
$H =$ transducer transfer function, and
$B(s) =$ output of transducer.

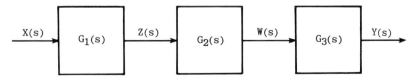

FIGURE D11. Block diagram for Figure D10.

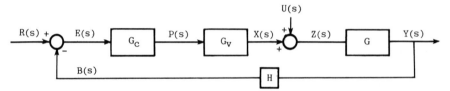

FIGURE D12. Closed-loop feedback control.

Using step-by-step block-diagram algebra,

$$Y(s) = GZ(s)$$
$$Z(s) = U(s) + X(s)$$
$$Y(s) = GU(s) + GX(s)$$
$$Y(s) = GU(s) + GG_v P(s)$$
$$Y(s) = GU(s) + GG_v G_c E(s)$$
$$Y(s) = GU(s) + GG_v G_c [R(s) - B(s)]$$
$$Y(s) = GU(s) + GG_v G_c R(s) - GG_v G_c B(s)$$
$$Y(s) = GU(s) + GG_v G_c R(s) - GG_v G_c H Y(s)$$

or

$$(1 + GG_v G_c H)Y(s) = GU(s) + GG_v G_c R(s)$$

or

$$Y(s) = \left(\frac{G}{1 + GG_v G_c H}\right)U(s) + \left(\frac{GG_v G_c}{1 + GG_v G_c H}\right)R(s)$$

These could have been written by inspection. Note that the numerator is the product of all transfer functions between the input and the output of the open-loop system. The denominator of all closed-loop systems is one plus the product of all transfer functions around the closed loop.

F.   Controllers

The usual industrial controller is pneumatic. They are sold with up to three different control modes.

1.   Proportional Controllers

The idealized transfer function for the proportional controller is

$$G_c = \frac{P(s)}{E(s)} = K_c$$

where K_c is a constant, usually called the gain.

2. Proportional-Derivative Controllers

The idealized transfer function for the proportional-derivative controller is

$$G_c = P(s)/E(s) = K_c(1 + \tau_D s)$$

where τ_D is the derivative time.

3. Proportional-Integral Controllers

The idealized transfer function for the proportional-integral controller is

$$G_c = \frac{P(s)}{E(s)} = K_c\left(1 + \frac{1}{\tau_I s}\right)$$

where τ_I is the integral time.

G. Stability

1. Introduction

If the output of a controlled system is bounded for all bounded inputs, the system is stable. Since most linear systems are exponential in time, that is,

$$y(t) = f(e^{-at})$$

if the exponent a is positive, the system is stable since $y(t)$ damps out as time gets large. If the exponent a were negative the system would be unstable since $y(t)$ would approach infinity as time increased. Only the simpler methods for determining stability will be discussed.

2. Partial Fraction Expansion

We have already seen that the Laplace transform function

$$Y(s) = \frac{A}{s + a}$$

gives the time function

$$Y(t) = Ae^{-at}$$

If we set the denominator of $Y(s)$ to zero,

$$s + a = 0$$

or

$$s = -a$$

we obtain the exponential term in $y(t)$. Were we to have a transfer function such as

$$Y(s) = \frac{A}{(s + s_1)(s + s_2)}$$

we would write it as

$$Y(s) = \frac{K_1}{s + s_1} + \frac{K_2}{s + s_2}$$

The time solution would be

$$Y(t) = K_1 e^{-s_1 t} + K_2 e^{-s_2 t}$$

and it would be stable if both s_1 and s_2 were positive. To find K_1 and K_2, we write $Y(s)$ with the common denominator

$$Y(s) = \frac{K_1(s + s_2) + K_2(s + s_1)}{(s + s_1)(s + s_2)}$$

Equate this numerator with the original numerator

$$K_1(s + s_2) + K_2(s + s_1) = A$$

or

$$(K_1 + K_2)s + (K_1 s_2 + K_2 s_1) = A$$

Equating coefficients of like powers of s on both sides of the equation gives

$$K_1 + K_2 = 0$$
$$K_1 s_2 + K_2 s_1 = A$$

or

$$K_1 = \frac{A}{s_2 - s_1}, \qquad K_2 = -\frac{A}{s_2 - s_1}$$

Thus the time solution is

$$y(t) = \left(\frac{A}{s_2 - s_1}\right)e^{-s_1 t} - \left(\frac{A}{s_2 - s_1}\right)e^{-s_2 t}$$

3. Characteristic Roots

Suppose a closed-loop system has the transfer function form

$$Y(s) = \frac{G_1}{1 + G_1 G_2} R(s)$$

We would manipulate the equation so that we have both a numerator and a denominator polynomial in the Laplace variable, s:

$$Y(s) = \frac{N(s)}{D(s)}$$

The denominator polynomial $D(s)$ is called the characteristic equation. Its roots [the values of s that make $D(s) = 0$] are called the characteristic roots. If we have any real, positive roots, we know that the system is unstable. Knowing the roots enables us to use the partial fraction expansion to obtain the time solution.

4. Routh's Method

When $D(s)$ is higher than a second-order polynomial in s, numerical root-finding procedures are necessary. But Routh's criterion allows us to determine stability without determining the roots.

We write the characteristic equation in the form

$$a_0 s^n + a_1 s^{n-1} + a_2 s^{n-2} + \cdots + a_{n-1} s + a_n = 0$$

where a_0 is positive. If any of the a_i are negative, the system is unstable. If all the a_i are positive, develop the Routh array:

Row					
1	a_0	a_2	a_4	a_6	$\cdots$
2	a_1	a_3	a_5	a_7	$\cdots$
3	b_1	b_2	b_3		
$\vdots$					
n	f_1				
$n+1$	g_1				

The elements in the third and succeeding rows are found from

$$b_1 = \frac{a_1 a_2 - a_0 a_3}{a_1}, \qquad b_2 = \frac{a_1 a_4 - a_0 a_5}{a_1}$$

$$c_1 = \frac{b_1 a_3 - a_1 b_2}{b_1}, \qquad c_2 = \frac{b_1 a_5 - a_1 b_3}{b_1}$$

The elements for the other rows are found from formulas which correspond to those given above. All elements of the first column of the Routh array must be positive and nonzero for a stable system.

PROBLEMS

A. A 70% NaOH liquor at 160°F with a heat content of 325 Btu/lb is diluted with a 20% NaOH liquor at 80°F with a heat content of 40 Btu/lb to give a final solution containing 40% NaOH. For every 100 lb of product, how much heat must be removed to maintain a final product temperature of 100°F? (Heat content of 40% solution is 94 Btu/lb at 100°F.)

B. A kiln is operated countercurrently to produce lime by burning a calcium carbonate sludge. The compositions of the flue gas leaving the cold end of the kiln and the sludge entering the same end are

Sludge	wt %	Flue Gas	Vol%
$CaCO_3$	45	CO_2	20
H_2O	50	CO	1
Inerts	5	N_2	76
		O_2	3

Methane, at a rate of 30,000 ft³/hr (measured dry at 60°F, 1 atm), is used to fire the kiln. If the lime conversion is 90% complete, how many pounds per hour of CaO is produced?

C. Benzene reacts with chlorine in a batch reactor to produce both mono- and dichlorobenzene according to

$$C_6H_6 + Cl_2 \longrightarrow C_6H_5Cl + HCl$$
$$C_6H_5Cl + Cl_2 \longrightarrow C_6H_4Cl_2 + HCl$$

A 1000-lb charge of benzene produced a product of 50 wt% monochlorobenzene, 30 wt% dichlorobenzene, and 20% unreacted benzene. The gas produced consisted of 90 vol% HCl and 10 vol% Cl_2 at 1 atm and 200°F. We wish to determine

(a) The volume of gas produced.
(b) The final weight of liquid product.
(c) The percentage of excess chlorine added, assuming the theoretical amount of chlorine needed would produce pure monochlorobenzene.

D. Air and gaseous HCl at 25°C are mixed and fed to a reactor in the Deacon process. The gaseous product leaves the reactor at 500°C with the composition:

Product	Vol%
HCl	12
N_2	52
O_2	6
Cl_2	14.5
H_2O	14.5

If the reaction stoichiometry is

$$4HCl + O_2 \longrightarrow 2Cl_2 + 2H_2O$$

how much heat must be supplied to or removed from the reactor for each 100 g mol of product gas? Assume no heat losses.

E. Given the control system of Figure D13 with $\tau_1 = 1$, $\tau_2 = \frac{1}{2}$, $\tau_3 = \frac{1}{3}$, determine the range of K_c for which the system is stable.

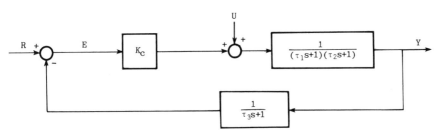

FIGURE D13. Schematic for problem E.

F. Given the block diagram of the control system shown in Figure D14, if Y is fed back positively, what K range is needed for stability? If Y is fed back negatively, what K range is needed for stability?

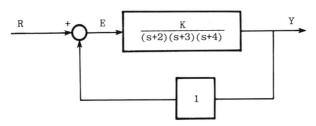

FIGURE D14. Schematic for Problem F.

G. What reaction temperature T maximizes the reversible reaction rate for

$$A \underset{k_2}{\overset{k_1}{\rightleftarrows}} B$$

if the concentrations are fixed and

$$k_i = \bar{k}_i e^{-E_i/RT}$$

with

$$E_2 > E_1$$

PROBLEM-SOLVING STRATEGIES

A. Using a basis of 100 lb of product, do a total material balance and an NaOH balance to determine the stream amounts. Then a heat balance should provide the answer.

B. This problem looks straightforward but involved. I need to deal with two reactions: the burning of methane with air and the decomposition of calcium carbonate. I must allow for incomplete combustion of methane, so we shall have two methane reactions. The flue gas CO_2 will come from the decomposition of lime and from the complete combustion of methane reactions. The N_2 comes directly from the air; the O_2 comes from excess air. If I had the time, I would work out the problem. I suspect that I would have to think carefully as I worked it out.

C. With this material balance problem, start with the stoichiometric equations, relating products to reactants. The problem would require logical thinking along the way to a solution. But I would not hesitate to attempt it.

D. With a basis of 100 g mol of product gas, determine the enthalpy of the product at 500°C. If I assume a base temperature of 25°C, I do not need to worry about the reactant enthalpy. Find the heats of combustion or formation of the components so that I can calculate the heat of reaction. Once I find the heat of reaction for each 2 mol of Cl_2, multiply by 14.5/2 to obtain the total heat. Finally, a heat balance provides the heat that must be added to the reactor. To solve this problem I need C_p data. If I can find it, mean C_p data are the most convenient to use.

E. Perform the block diagram algebra, step by step, to get Y as a function of R. Find the denominator polynomial in s. Use the Routh criterion.

F. Perform the block diagram with a in the 1 block. Find Y as a function of R and a. Get the denominator polynomial in s (and a). Use the Routh criterion for stability. When $a = +1$, the answer is for the first question; when $a = -1$, the second question is answered.

G. Set up the equation for the rate of formation of B in terms of k_1, k_2, C_a, and C_b. Find the stationary point by differentiating r_b with respect to T. Substitute this value of T into the r_b equation for the solution. Make sure that the second derivative of r_b with respect to T is negative; this assures a maximum.

SOLUTIONS

A. See Figure D15.
Basis: 100 lb of 40% NaOH solution at 100°F.

$$\text{Total material balance:} \quad X + Y = 100$$
$$\text{NaOH balance:} \quad 0.70X + 0.20Y = 40$$
$$\text{Enthalpy balance:} \quad 325X + 40Y = Q + 94(100)$$

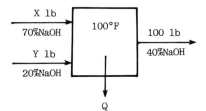

FIGURE D15. Schematic for solution A.

So $X = 100 - Y$.

Then $0.70(100 - Y) + 0.20Y = 40$ or $70 - 0.50Y = 40$

or $0.50Y = 30$ or $Y = 60$

Then $X = 40$

Using enthalpy balance,

$$325(40) + 40(60) = Q + 9400$$

or

$$Q = 6000 \text{ Btu/100 lb of product solution}$$

B. See Figure D16.

Basis: 100-lb sludge

$$\begin{array}{ccccc} & CaCO_3 & = & CaO & + & CO_2 \\ MW & (100.0) & & (56) & & (44) \end{array}$$

Sludge	Wt%	Pounds	Moles
$CaCO_3$	45	45	0.45
H_2O	50	50	
Inerts	5	5	
	100	100	

moles CaO made $= 0.90(0.45) = 0.405$

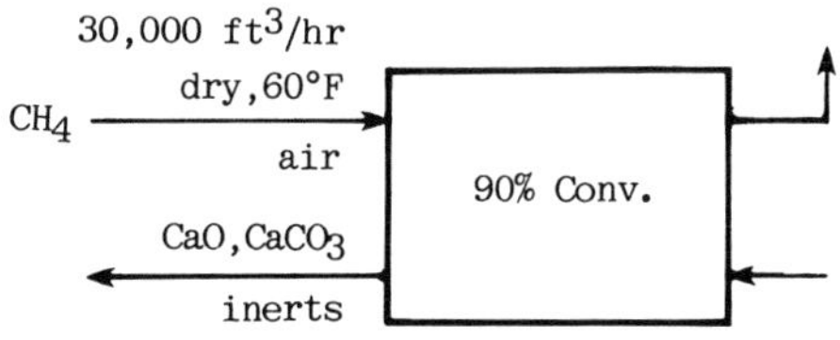

FIGURE D16. Schematic for solution B.

The combustion reactions:

$$CH_4 + 2O_2 = CO_2 + 2H_2O$$
$$CH_4 + \tfrac{3}{2}O_2 = CO + 2H_2O$$

So the flue gas CO_2 comes from both the $CaCO_3$ reaction and the CH_4 reaction.

Assume Z moles of flue gas per 100-lb sludge:

Flue Gas	Mole %	Moles
CO_2	20	0.20 Z
CO	1	0.01 Z
N_2	76	0.76 Z
O_2	3	0.03 Z
	100	Z

Moles O_2 in air used for burning $= (0.76\,Z/0.79)0.21 = 0.202\,Z$.
Moles O_2 in combustion $= 0.202\,Z - 0.03\,Z = 0.172\,Z$.
Moles CO_2 from combustion $= 0.20\,Z - 0.405$.
Moles O_2 burning CH_4 to $CO_2 = 2(0.20\,Z - 0.405) = 0.4\,Z - 0.81$.
Moles O_2 burning CH_4 to $CO = 0.172\,Z - (0.4\,Z - 0.81)$
$$= 0.81 - 0.228\,Z.$$

Moles CO formed $= \tfrac{2}{3}(0.81 - 0.228\,Z) = 0.54 - 0.152\,Z = 0.01\,Z$
$0.142\,Z = 0.54 \qquad Z = 3.80 \text{ mol}$
Moles C from $CH_4 = (0.20\,Z + 0.01\,Z) - 0.405$
$$= 0.393 \text{ mol}$$

To get the pounds per hour of CaO produced,

$$CH_4 = 30{,}000 \frac{ft^3}{hr} \frac{492°R}{520°R} \frac{mol}{359\ ft^3} = 79.07 \frac{mol}{hr}$$

Then

$$\text{CaO production} = \frac{0.405 \text{ mol CaO}}{100 \text{ lb sludge}} \frac{56 \text{ lb CaO}}{\text{mol CaO}} \frac{100 \text{ lb sludge}}{0.93 \text{ mol } CH_4} \times$$

$$\frac{79.07 \text{ mol } CH_4}{hr} = 4563 \frac{lb}{hr}$$

C. *Basis*: 100-lb liquid product.

$$C_6H_6 + Cl_2 \longrightarrow C_6H_5Cl + HCl$$
MW (78.1)   (70.9)   (112.6)   (36.5)

$$C_6H_5Cl + Cl_2 \longrightarrow C_6H_4Cl_2 + HCl$$
MW (112.6)   (70.9)   (147.0)   (36.5)

Liquid	Wt%	Pounds	Moles	H	Moles Cl_2
C_6H_5Cl	50	50	$0.444 \times 5 = 2.220$		0.222
$C_6H_4Cl_2$	30	30	$0.204 \times 4 = 0.816$		0.204
C_6H_6	20	20	$0.256 \times 6 = 1.536$		
			0.904	4.572	0.426

moles C_6H_6/100-lb liquid product = 0.904 mol
lb atoms H in feed = 6(0.904) = 5.424
lb atoms H in gas = 5.424 − 4.572 = 0.852

Gas	Mole %	Moles	Moles Cl_2
HCl	90	0.852	0.426
Cl_2	10	0.095	0.095
	100	0.947	0.521

(a) Volume of gas produced $= \dfrac{0.947 \text{ mol gas}}{100 \text{ lb liq prod}} \dfrac{359 \text{ ft}^3}{\text{mol gas STP}}$

$\dfrac{100 \text{ lb liq prod}}{0.904 \text{ mol feed}} \dfrac{1000 \text{ lb } C_6H_6 \text{ feed}}{78.1 \text{ lb } C_6H_6/\text{mol } C_6H_6} = 4815 \text{ ft}^3 \text{ at STP}$

(b) Weight of liq prod $= \dfrac{100 \text{ lb liq prod}}{0.904 \text{ mol feed}} \times \dfrac{1000 \text{ lb feed}}{78.1 \text{ lb/mol}} = 1416 \text{ lb}$

(c) Theoretical $Cl_2 = 0.904$; total Cl_2 feed $= 0.426 + 0.521 = 0.947$

percentage excess $= \dfrac{0.947 - 0.904}{0.904} \, 100 = 4.8\%$

D. See Figure D17.
Basis: 100-g mol product gas
Base temperature 25°C.

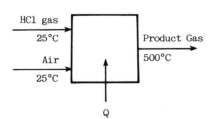

FIGURE D17. Schematic for solution D.

Vol % or

Product	Vol % or Mol %	Moles	Mean C_p at 25–500°C	Moles $\times C_p$(500–25)
HCl	12	12	[†]7.1 cal/g mol°C	40,470
N_2	53	53	7.2	181,260
O_2	6	6	7.5	21,375
Cl_2	14.5	14.5	8.6	59,233
H_2	14.5	14.5	8.5	58,544
				360,882 cal

Heats of formation at 25°C.

$$HCl(g) = -22,063 \text{ cal/g mol}^{[‡]}$$
$$O_2 = 0$$
$$Cl_2 = 0$$
$$H_2O(g) = -57,798 \text{ cal/g mol}$$

$$\Delta H_R 25°C = \Delta H_f(\text{prod}) - \Delta H_f(\text{react})$$
$$= 2(-57,798) - 4(-22,060) = -27,344 \text{ cal/2 g mol } Cl_2$$
$$= \frac{-27,344 \text{ cal}}{2 \text{ g mol } Cl_2} \times 14.5 \text{ g mol } Cl_2 = -198,244 \text{ cal}$$

$$Q - (-198,244) = 360,882 \text{ or}$$

$$Q = 162,638 \text{ cal supplied/100 g mol product}$$

E.

$$Y = K_c G_1 E \qquad G_1 = \frac{1}{(s+1)(\frac{1}{2}s+1)}$$

$$E = R - HY \qquad H = \frac{1}{\frac{1}{3}s+1}$$

$$Y = K_c G_1 (R - HY)$$

$$Y(1 + K_c G_1 H) = K_c G_1 R$$

Characteristic equation is numerator of

$$1 + K_c GH$$

[†] Smith and Van Ness, *Introduction to Chemical Engineering Thermodynamics*, McGraw-Hill, New York, 19 , 2nd ed., p. 126.
[‡] Same reference, p. 138.

or

$$1 + \frac{K_c}{(s + 1)(\frac{1}{2}s + 1)(\frac{1}{3}s + 1)}$$

The numerator is

$$(s + 1)(\tfrac{1}{2}s + 1)(\tfrac{1}{3}s + 1) + K_c$$
$$(\tfrac{1}{2}s^2 + \tfrac{3}{2}s + 1)(\tfrac{1}{3}s + 1) + K_c$$
$$\tfrac{1}{6}s^3 + \tfrac{1}{2}s^2 + \tfrac{1}{2}s^2 + \tfrac{3}{2}s + \tfrac{1}{3}s + 1 + K_c$$
$$\tfrac{1}{6}s^3 + s^2 + \tfrac{11}{6}s + 1 + K_c$$

or

$$s^3 + 6s^2 + 11s + 6(1 + K_c)$$

Routh array

$$
\begin{array}{cc}
1 & 11 \\
6 & 6(1 + K_c) \\
\dfrac{6(11) - 6(1 + K_c)}{6} & \\
6(1 + K_c) &
\end{array}
$$

All of first column must be positive: 1 is $(+)$, 6 is $(+)$, $11 - 1 - K_c$ must be $(+)$, $1 + K_c$ must be $(+)$.

$$K_c < 10 \qquad \text{and} \qquad K_c > -1$$

for stability

F.

$$Y = GE \qquad \text{positive feedback } a = 1$$
$$E = R + aY \qquad \text{negative feedback } a = -1$$
$$Y = G(R + aY) \qquad \text{or} \qquad Y(1 - aG) = GR$$

Characteristic equation is numerator of $1 - aG$

$$1 - aG = 1 - \frac{aK}{(s + 2)(s + 3)(s + 4)}$$

$$
\begin{aligned}
\text{numerator} &= (s + 2)(s + 3)(s + 4) - aK \\
&= (s^2 + 5s + 6)(s + 4) - aK \\
&= s^3 + 4s^2 + 5s^2 + 20s + 6s + 24 - aK \\
&= s^3 + 9s^2 + 26s + 24 - aK
\end{aligned}
$$

Routh array

$$
\begin{array}{cc}
1 & 26 \\
9 & 24 - aK \\
(9 \cdot 26 - 24 + aK)/9 & \\
24 - aK &
\end{array}
$$

First column 1, 9, $(210 + aK)/9$, $24 - aK$ must all be positive:
  For positive feedback $a = 1$ and $210 + K > 0$
and

$$24 - K > 0 \quad \text{or} \quad \text{for stability} \ -210 < K < 24$$

For negative feedback $a = -1$ and $210 - K > 0$
and

$$24 + K > 0 \quad \text{or} \quad \text{for stability} \ -24 < K < 210$$

G.
$$A \ \underset{k_2}{\overset{k_1}{\rightleftharpoons}} \ B$$

$$r_b = +k_1 C_a - k_2 C_b$$

$$\max (r_b) = \max (k_1 G_a - k_2 C_b)$$

$$\frac{dr_b}{dT} = C_a \frac{dk_1}{dT} - C_b \frac{dk_2}{dT}$$

$$k_1 = \bar{k}_1 e^{-E_1/RT}, \qquad k_2 = \bar{k}_2 e^{-E_2/RT}$$

$$\frac{dk_1}{dT} = e^{-E_1/RT} \frac{d}{dT} \left(-\frac{E_1}{R} T^{-1} \right) = \frac{E_1}{RT^2} e^{-E_1/RT}$$

$$\frac{dk_2}{dT} = \frac{E_2}{RT^2} e^{-E_2/RT}$$

$$\frac{d^2 r_b}{dT^2} = C_a \frac{d^2 k_1}{dT^2} - C_b \frac{d^2 k_2}{dT^2}$$

$$\frac{d^2 k_1}{dT^2} = \frac{E_1}{R} \left(\frac{1}{T^2} \frac{d}{dT} e^{-E_1/RT} - \frac{2}{T^3} e^{-E_1/RT} \right)$$

$$= \frac{E_1}{R} \left(\frac{E_1}{RT^4} e^{-E_1/RT} - \frac{2}{T^3} e^{-E_1/RT} \right)$$

Stationary point

$$C_a \frac{E_1}{RT^2} e^{-E_1/RT} = C_b \frac{E_2}{RT^2} e^{-E_2/RT}$$

$$\frac{C_a E_1}{C_b E_2} = e^{-(E_2 - E_1)/RT}$$

$$\frac{d^2 r_b}{dT^2} = \frac{E_1}{R^2 T^4} e^{-E_1/RT}(E_1 - 2RT)C_a - \frac{E_2}{R^2} e^{-E_2/RT}(E_2 - 2RT)C_b$$

$$= \frac{C_b E_2}{R^2 T^4} e^{-E_2/RT}(E_1 - 2RT) - \frac{C_b E_2}{R^2 T^4} e^{-E_2/RT}(E_2 - 2RT)$$

$$= \frac{C_b E_2}{R^2 T^4} e^{-E_2/RT}(E_1 - 2RT - E_2 + 2RT)$$

$$= \frac{C_b E_2 (E_1 - E_2)}{R^2 T^4} e^{-E_2/RT}$$

This is negative so the stationary point gives a maximum.

MASS TRANSFER

W. L. McCabe and J. C. Smith, *Unit Operations of Chemical Engineering*, 3rd ed. McGraw-Hill, New York, 1976.

I. INTRODUCTION

A. Topics Covered

We shall discuss the general stage principles and then some of the principal methods of mass transfer: distillation, leaching and extraction, absorption, humidification, and drying. We shall discuss only the material that is likely to appear on the P.E. exam. Therefore, little on multicomponent separations will be found here. The theoretical diffusion and mass transfer between phases will not be discussed; the topic is amply covered by McCabe and Smith.

B. Symbols

McCabe and Smith's convention for identifying the vapor V phase and the liquid L phase will be used:

Process	V Phase	L Phase
Distillation	Vapor	Liquid
Gas absorption	Gas	Liquid
Liquid extraction	Extract	Raffinate
Leaching	Liquid	Solid
Humidification	Gas	Liquid
Drying	Gas	Wet solid

The subscript a will refer to the end of the unit where the L phase enters; b to the end where it leaves.

C. Stage Principles

1. Terminology

Stages are numbered in the direction of flow of the L phase, with the nth stage being a general stage. The total number of stages is N.

2. Material Balances

Doing a total material balance into and out of envelope 1 of Figure M1 gives

$$L_a + V_{n+1} = L_n + V_a$$

for steady-state operation. A balance on component A of a two-component system gives

$$L_a x_a + V_{n+1} y_{n+1} = L_n x_n + V_a y_a$$

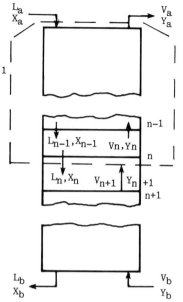

FIGURE M1. Terminology for a staged process.

Had we chosen to write the balances over the whole unit, we would obtain

$$L_a + V_b = L_b + V_a$$

and

$$L_a x_a + V_b y_b L_b x_b + V_a y_a$$

3. Enthalpy Balances

If the enthalpies are denoted by H, the envelope 1 enthalpy balance is

$$L_a H_{L,a} + V_{n+1} H_{V,n+1} = L_n H_{L,n} + V_a H_{V,a}$$

while the enthalpy balance around the total unit is

$$L_a H_{L,a} + V_b H_{V,b} = L_b H_{L,b} + V_a H_{V,a}$$

4. Graphical Method

Many two-component problems can be solved graphically. If either the operating line or the equilibrium is straight, graphical methods are ideal. If both are straight, the various equations are quicker and easier.

a. The Operating Line

The operating line is simply the component material balance expressed as $y =$ function of x:

$$y_{n+1} = \frac{L_n}{V_{n+1}} x_n + \frac{V_a y_a - L_a x_a}{V_{n+1}}$$

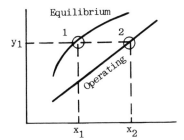

FIGURE M2. Graph of an ideal equilibrium stage.

Note that the two end points, (x_a, y_a) and (x_b, y_b), satisfy this material balance equation. In fact, all stages in between must also be represented by this material balance equation. Of course, it is much easier to use when it is a straight line; that is, when L and V are constant or when they can be redefined so as to be constant.

b.   Ideal Stage (or Perfect Plate)

We live with actual stages, but it is easier to calculate the theoretical or ideal stages. With the number of ideal stages in hand, we can apply a stage efficiency factor to obtain the actual stages.

The ideal stage is defined such that the V phase leaving a stage is in equilibrium with the L phase leaving the same stage. Thus we need the equilibrium curve relating y to x. As shown in Figure M2, the two phases leaving the same stage are related by the equilibrium curve (point 1), and the phases from one stage to the next are related by the operating line (material balance), or point 2. Continuing up or down the unit in this manner allows one to determine the number of stages needed to effect the separation.

5.   Absorption Factor Method

If both the equilibrium line and the operating line are straight, the graphical stage-to-stage construction can be expressed as a formula. We assume the straight equilibrium line to be

$$y_e = mx_e + B$$

or, for the ideal, nth stage,

$$y_n = mx_n + B$$

Substituting this into the operating line (straight when we assume L/V constant),

$$y_{n+1} = \frac{L(y_n - B)}{mV} + y_a - \frac{Lx_a}{V}$$

Define the absorption factor $A = L/(mV)$, so

$$y_{n+1} = Ay_n - A(mx_a + B) + y_a$$

Now note that the $(mx_a + B)$ quantity is the vapor concentration that is in equilibrium with the inlet L phase concentration x_a. Denote it by $y_a^* = mx_a + B$, so that

$$y_{n+1} = Ay_n - Ay_a^* + y_a$$

Now do the stage-by-stage construction with $n = 1$, or

$$y_2 = Ay_1 - Ay_a^* + y_a = Ay_a - Ay_a^* + y_a$$
$$= y_a(1 + A) - Ay_a^*$$

since $y_a = y_1$. Then for $n = 2$

$$y_3 = Ay_2 - Ay_a^* + y_a = A[y_a(1 + A) - Ay_a^*] - Ay_a^* + y_a$$
$$= y_a(1 + A + A^2) - y_a^*(A + A^2)$$

This eventually becomes

$$y_b = y_a \frac{1 - A^{N+1}}{1 - A} - y_a^* A \frac{1 - A^N}{1 - A}$$

This is the Kremser equation. It can be solved for N:

$$N = \ln\left\{\frac{y_b - y_b^*}{y_a - y_a^*}\right\} \bigg/ \ln A$$

Our equation can also be expressed in terms of L-phase concentrations x, and the stripping factor $S = 1/A = mL/V$, or

$$N = \ln\left\{\frac{x_b - x_b^*}{x_a - x_a^*}\right\} \bigg/ \ln S$$

6. Enthalpy-Concentration Diagram Method

We have an A stream at a rate $\dot{m}_a$ mixed with a B stream of rate $\dot{m}_b$ to give a C stream. With concentrations x_a, x_b, x_c and enthalpies H_a, H_b, H_c, we can show that the enthalpy balance is

$$\dot{m}_a H_a + \dot{m}_b H_b = (\dot{m}_a + \dot{m}_b)H_c$$

and the component material balance is

$$\dot{m}_a x_a + \dot{m}_b x_b = (\dot{m}_a + \dot{m}_b)x_c$$

Thus

$$\frac{\dot{m}_a}{\dot{m}_b} = \frac{H_b - H_c}{H_c - H_a} = \frac{x_b - x_c}{x_c - x_a}$$

the lever principle. McCabe and Smith (pp. 528–530) show how this may be modified for nonadiabatic processes.

II. DISTILLATION

A. Flash Distillation

Consider a liquid mixture in which the pressure is reduced (usually through a valve). A vapor–liquid equilibrium results, and the two phases are separated. If we have 1 mol of a binary mixture with x_f mole fraction of the more volatile component and we let f be the fraction of the feed vaporized, then $1 - f$ is the fraction of feed that leaves as liquid. If y_d and x_b are the vapor and liquid concentrations, then a more volatile component balance becomes

$$x_f = fy_d + (1 - f)x_b$$

We know neither y_d nor x_b, but we have assumed that equilibrium exists. So, once again, we need the equilibrium line to go with our material balance line. With the equilibrium assumption, the point (y_d, x_b) falls on the equilibrium curve, y = function of x, so we shall write the material balance equation without the subscripts

$$x_f = yf + (1 - f)x$$

or

$$y = -\left(\frac{1-f}{f}\right)x + \frac{x_f}{f}$$

which is a straight line with slope $-(1 - f)/f = -L/V$ passing through the $y = x$ diagonal at $x = x_f$.

B. Differential Distillation

Now consider a liquid charge of n_0 moles initially in a still. At any time t, there will be n moles of liquid remaining in the still with a liquid composition x. At this same time t, the vapor composition is y, and the amount of component A will be $n_a = xn$.

At this instant, we assume that a small amount of liquid is vaporized, dn moles. The amount of A in this vapor will be $-dn_a$ or $-y\,dn$, since n_a is decreasing. Thus

$$dn_a = d(xn) = x\,dn + n\,dx$$

so

$$y\,dn = x\,dn + n\,dx$$

or

$$\frac{dn}{n} = \frac{dx}{y - x}$$

This equation will be integrated from the initial charge n_0 to the final charge n_1, while the x goes from x_0, the initial concentration to x_1, the final concentration, giving the Rayleigh equation:

$$\int_{n_1}^{n_0} \frac{dn}{n} = \ln\left(\frac{n_0}{n_1}\right) = \int_{x_1}^{x_0} \frac{dx}{y - x}$$

This equation may be integrated graphically using the equilibrium curve, $y =$ function of x, or if the relative volatility

$$\alpha_{ab} = \frac{y_a}{x_a} \frac{x_b}{y_b} = \frac{y}{1 - y} \frac{1 - x}{x}$$

is constant, the equation may be integrated to

$$\frac{n_b}{n_{b0}} = \left(\frac{n_a}{n_{a0}}\right)^{1/\alpha_{ab}}$$

where subscripts a and b refer to these components and the subscript 0 refers to the initial charge.

C.   Steam Distillation

Many organics are temperature sensitive; conventional distillation procedures would be inappropriate. For these cases, steam distillation may deserve consideration. Here, volatile organic liquids are separated from nonvolatile impurities by vaporization of the volatiles by passing live steam through them. The general equation for this scheme is

$$\frac{\dot{m}_s}{\dot{m}_a} = \frac{M_s}{M_a} \cdot \frac{p_s}{p_a}$$

where $\dot{m}_s$, $\dot{m}_a$ = mass flow rates of vapor, steam and volatile,
 M_s, M_a = molecular weights, steam and volatile, and
 p_s, p_a = partial pressures, steam and volatile.

This equation reduces, for specific cases, to:
    If the steam does not condense,

$$\frac{\dot{m}_s}{\dot{m}_a} = \left(\frac{M_s}{M_a}\right) \cdot \frac{\pi - VP_a}{VP_a}$$

where π = total pressure and
 VP_a = vapor pressure of volatile.

If the steam condenses and is immiscible (forming two liquid layers),

$$\frac{\dot{m}_s}{\dot{m}_a} = \frac{M_s}{M_a} \frac{VP_s}{VP_a} = \frac{M_s}{M_a} \cdot \frac{\pi - VP_a}{VP_a}$$

If the steam condenses and is miscible (forming but one liquid layer),

$$\frac{\dot{m}_s}{\dot{m}_a} = \frac{M_s \, x_s \cdot VP_s}{M_a \, x_a \cdot VP_a}$$

where x_s, x_a = mole fraction in liquid, steam and volatile

D.   Continuous Distillation

1.   Principles

In the stage principles section we obtained the material balance or operating line. For distillation, we have the same operating line form, but with the distillation notation shown in Figure M3. Now we have two, one above the feed and one below the feed:

$$y_{n+1} = \frac{L_n}{V_{n+1}} \, x_n + \frac{Dx_d}{V_{n+1}} \qquad \text{(above feed)}$$

$$y_{m+1} = \frac{L_m}{V_{m+1}} \, x_m - \frac{Bx_b}{V_{m+1}} \qquad \text{(below feed)}$$

Using a total balance around the top of the column, we find

$$D = V_{n+1} - L_n$$

and around the bottom

$$B = L_m - V_{m+1}$$

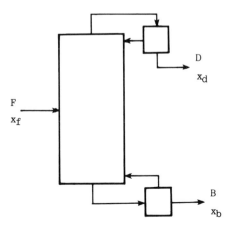

FIGURE M3.   A distillation column.

Eliminating V_{n+1} and V_{m+1} we obtain for the material balances or operating lines

$$y_{n+1} = \frac{L_n}{L_n + D} x_n + \frac{Dx_d}{L_n + D}$$

and

$$y_{m+1} = \frac{L_m}{L_m - B} x_n - \frac{Bx_b}{L_m - B}$$

2. McCabe–Thiele Graphical Method

If we can assume constant molal overflow, that is L_n and L_m are constant, the operating lines are straight. Assuming that we have available the equilibrium data for our binary, we may use the McCabe-Thiele graphical method.

Most analyses use the concept of reflux ratio or R_D,

$$R_D = \frac{L}{D} = \frac{V - D}{D}$$

With this definition, the above feed operating line is

$$y_{n+1} = \frac{R_D}{R_D + 1} x_n + \frac{x_d}{R_D + 1}$$

Note that the operating line passes through the $y = x$ diagonal at $x = x_d$. This is true regardless of whether we are using a total or a partial condenser. But keep in mind that if a partial condenser is being used, we subtract one theoretical stage when counting stages. Thus we see that the operating line for the rectifying section (above the feed) goes through the point $y = x = x_d$ with a slope of $R_D/(R_D + 1)$.

The operating line for the stripping section (below the feed) was

$$y = \frac{\bar{L}}{\bar{L} - B} x - \frac{Bx_b}{\bar{L} - B}$$

with $\bar{L}$ the liquid flow down the column beneath the feed point. You will notice that this equation goes through the point $y = x = x_b$, regardless of the type of reboiler. Bear in mind though that a partial reboiler is a theoretical stage. Thus, in counting the theoretical trays or plates required, you must subtract one for a partial reboiler.

The feed plate separates the stripping section from the rectifying section. The addition of the feed increases the reflux in the stripping section, increases the vapor in the rectifying section, or does both.

For example, if the feed is cold, the entire feed stream becomes a part of the stripping section reflux. But also, to heat the feed to its bubble point, some vapor must condense, and this condensate also becomes part of the stripping section reflux. Of course the vapor to the rectifying section is less, by the amount of the condensate.

To determine numerically the effect of the feed on the column internal flows, we use the q factor, which is defined as the moles of liquid flow in the stripping section that result from each mole of feed. Equivalently, it could be defined as the heat required to convert one mole of feed to saturated vapor divided by the molal heat of vaporization of the feed. The various possibilities are

$q > 1$, cold feed,

$q = 1$, feed at bubble point (saturated liquid),

$0 < q < 1$, feed partially vapor,

$q = 0$, feed at dewpoint (saturated vapor), and

$q < 0$, feed superheated vapor.

Note that if the feed is a mixture of liquid and vapor, q is the fraction that is liquid.

With the definition of q, the feed line can be derived as

$$ y = \frac{q}{q - 1} x - \frac{x_f}{q - 1} $$

Once again, the line passes through the diagonal at $x = x_f$, that is, $y = x = x_f$ and the slope of the line is $q/(q - 1)$. All three lines, the rectifying, the feed, and the stripping pass through a common point.

Once the rectifying and stripping operating lines have been plotted, we start stepping in the stages. But when are we at the feed stage? When we change from stripping to rectifying or from rectifying to stripping. Note that this is *not* unique. But we get the smallest number of stages by changing as soon as it is possible.

What is the minimum number of stages that we could possibly use to obtain a desired separation? The more reflux we use, the better the separation. But if the separation is fixed, the more reflux we use, the fewer stages we need. So, in the limit, an infinite reflux ratio should give us the minimum stages. And an infinite reflux ratio gives us a slope of one, or the diagonal line. Thus, using the $y = x$ diagonal as the operating line and stepping off the stages determines the minimum number of stages.

At the other extreme, as the reflux ratio gets smaller, the number of stages required increases, to a point where we require an infinite number of stages. Graphical analysis will show that if the operating line passes through the intersection of the feed line and the equilibrium line, an infinite number of stages would occur at the pinch point.

The slope of this operating line is $R_{min}/(R_{min} + 1)$, from which we can calculate the minimum reflux ratio R_{min}. Should the equilibrium line be so nonideal that the operating line passes through it, we increase the slope of the operating line until it just becomes tangent to the equilibrium line. The reflux ratio calculated for this slope is then the minimum.

3. *Ponchon–Savarit Enthalpy-Concentration Method*

This method is fully discussed by McCabe and Smith, pp. 571–585.

III. LEACHING AND EXTRACTION

We now discuss methods of removing one constituent from a solid or liquid by means of a liquid solvent. There are two categories. Liquid extraction is used to separate two miscible liquids with the use of a solvent which preferentially dissolves one of them. Leaching (solid extraction) is used to dissolve soluble matter from its mixture with an insoluble solid.

A. Leaching

1. *Terminology*

The most important method is the continuous countercurrent method using stages. Consider the diagram in Figure M4 where the stages are numbered in the direction of the solids flow. The V phase is liquid that overflows from stage to stage counter to that of the solid, dissolving solute as it moves from stage N to stage 1. The L phase is the solid. Exhausted solids leave stage N and concentrated solution overflows from stage 1.

We assume that the solute-free solid is insoluble in the solvent and that the flow rate of this solid is constant throughout the cascade. We let V refer to the overflow solution and L to the liquid retained by the solid, both based on a definite flow of dry solute-free solid. The terminal concentrations are $x_a =$ solution on entering solid, $x_b =$ solution on leaving solid, $y_b =$ fresh solvent entering the system and $y_a =$ concentrated solution leaving the system. Just as with distillation or absorption, the calculations for a countercurrent system can be analyzed by using an equilibrium line and an operating line. The method used depends upon whether these lines are straight or curved.

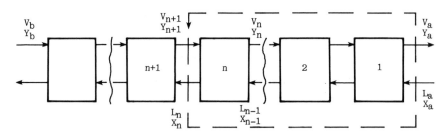

FIGURE M4.   A staged leaching system.

2. Equilibrium

If sufficient solvent is present to dissolve all the solute in the entering solid, equilibrium is attained. And the concentration of the liquid retained by the solid leaving any stage is the same as that of the liquid overflow from the stage. The equilibrium relationship is $x_e = y_e$.

3. Operating Line

The operating line equation is obtained from material balances around the first n stages:

total solution: $V_{n+1} + L_a = V_a + L_n$

solute: $V_{n+1}y_{n+1} + L_a x_a = L_n x_n + V_a y_a$

Then

$$y_{n+1} = \frac{1}{1 + [(V_a - L_a)/L_n]} x_n + \frac{V_a y_a - L_a x_a}{L_n + V_a - L_a}$$

This operating line passes through the points $(x_a, y_a), (x_b, y_b)$.

4. Constant or Variable Underflow

Two cases are considered. If solution density and viscosity change considerably with solute concentration, the lower stages may have solids with more liquid retention than the higher stages. The slope of the operating line would thus vary from stage to stage. On the other hand, if the mass of solution retained by the solid is independent of concentration, L_n is constant and the operating line is straight. This is called constant solution underflow. Of course if the underflow is constant, so is the overflow.

5. Number of Ideal Stages—Constant Underflow

When the operating line is straight, a McCabe–Thiele diagram can be used to determine the number of ideal stages; but with leaching, the equilibrium line is always straight, so the Kremser equation

$$N = \ln\left(\frac{y_b - y_b^*}{y_a - y_a^*}\right)\Big/\ln\left(\frac{y_b - y_a}{y_b^* - y_a^*}\right)$$

can be used with $y_a^* = x_a$ and $y_b^* = x_b$.

Note that this equation cannot be used for the entire cascade if L_a, the solution entering with the unextracted solids, differs from L, the underflow within the system. This is illustrated later in the problems.

6. Number of Ideal Stages—Variable Underflow

When the underflow and overflow vary stage to stage, a modification of Ponchon–Savarit can be used. The modifications are (1) consider each stream a mixture of solid and solution and (2) use the ratio of solid to solution in place of enthalpy. The solution is a mixture of solute and solvent. Let a represent solute, b the solid, and S the solvent. Then the abscissa is $X = a/(a + s)$. The ordinate is $Y = b/(a + s)$. The curved line of Figure M5 must usually be determined experimentally. Since the solid is assumed insoluble in the solvent, all points representing overflow streams lie upon the X axis ($Y = 0$). Also since at equilibrium, the solution concentrations for both under- and over-flow for the same stage are equal, all tie lines are vertical. As with Ponchon–Savarit, a point for the mixture of two streams lies on the straight line connecting the mixed streams and the level rule applies. Bear in mind that lines on the X-Y diagram represent not the masses of the total streams but only the masses of the solution in the streams.

a. Overall Balances and Point J

The overall balance for leaching has four streams: the inlet solvent V_b, the outlet concentrated liquid V_a, the inlet feed L_a, and the outlet exhausted solid L_b. We assume the overall process replaced by two fictitious processes in series; in the first, streams V_b and L_a are mixed to form stream J; in the second, stream J is divided into streams V_a and L_b. These processes are represented in Figure M6 by

$$V_b + L_a = J = L_b + V_a$$

Five points, V_b, L_a, J, V_a, L_b, each representing a corresponding stream, are shown. A straight line through points V_a and L_b intersects a straight line through points V_b and L_a at point J. By the lever rule, the mass of solution in stream L_a is inversely proportional to the line $\overline{JL_a}$; the solution in stream V_a is inversely

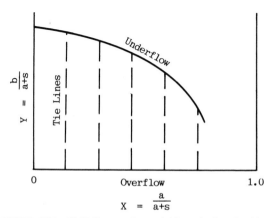

FIGURE M5. X-Y diagram for variable underflow leaching.

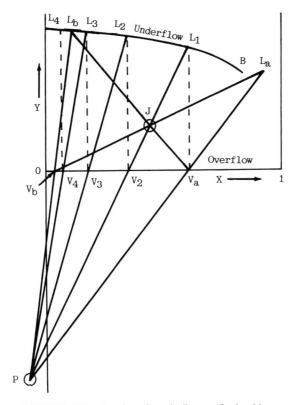

FIGURE M6. Ponchon–Savarit diagram for leaching.

proportional to line $\overline{JV_a}$; and so forth. If, for example, the mass ratio of the solution in the incoming solvent to that in the feed is known, this ratio equals $\overline{JL_a}/\overline{JV_a}$ and point J can be plotted. Then if the analysis of one other stream, for example, V_a, is known, the line $L_b J V_a$ can be plotted. Since point L_b is on the underflow line, it is also the intersection of the underflow line with line $L_b J V_a$, and L_b is thus determined.

b. Operating Line and Point P

Overall balances are treated by assuming a fictitious mixer added to each end of the cascade. Imaginary stream P is assumed to be formed by subtracting L_a from V_a or by subtracting L_b from V_b:

$$V_a = L_a + P; \quad L_b + P = V_b \quad \text{or} \quad P = V_a - L_a = V_b - L_b$$

Then a material balance over the first n stages and the fictitious mixer ahead of stage 1 is

$$V_{n+1} = L_n + P$$

The operating line represented is a straight line passing through points P, V_{n+1}, and L_n. All other individual operating lines also pass through point P. These lines, used alternately with the vertical equilibrium lines, give the Ponchon–Savarit construction shown.

B. Liquid Extraction

When distillation is difficult or ineffective, liquid extraction is used. Close-boiling mixtures are prime candidates for extraction. This method uses differences in the solubilities of the components, thus exploiting chemical differences.

1. Basic Data

(a) A and B are substantially insoluble liquids, C is a distributed solute.

(b) $E = $ lb/hr ft^2 extract (solvent rich)
$R = $ lb/hr ft^2 raffinate (residual liquid)
$B = $ lb/hr ft^2 solvent

(c) $x = $ wt   fraction of C in raffinate
$y = $ wt   fraction of C in extract
$X = x/(1 - x)$ lb C/lb non-C in raffinate
$Y = y/(1 - y)$ lb C/lb non-C in extract

(d) Triangular coordinates. The sum of the perpendicular distance from any point within the triangle to the three sides equals the altitude of the triangle, as shown in Figure M7. Let altitude $= 100\%$. Point D: binary 20% B, 80% A. All points on DC have same ratio of A to B.

Adding R pounds of mixture at point R to E pounds of mixture at E gives the new mixture shown on the straight line at point M, such that

$$\frac{R}{E} = \frac{\overline{ME}}{\overline{RM}} = \frac{X_E - X_M}{X_M - X_R} \text{ mixture rule}$$

(e) Kinds of Systems

   (1) Type I (see Figure M8). Three liquids, one pair partially soluble; C dissolves completely in A and B; A and B soluble to a limited ex-

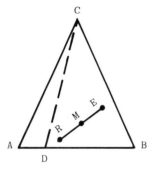

FIGURE M7. Triangular coordinates.

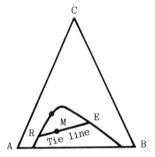

FIGURE M8. Extraction equilibrium diagram—type I.

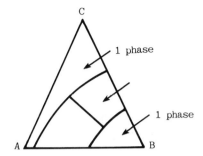

FIGURE M9. Extraction equilibrium diagram—Type II.

tent. (More C in E than in R—distribution coefficient greater than 1: large distribution Coefficient—less extracting solvent required.)

(2) Type II (see Figure M9). A and C completely soluble; A and B, B and C limited solubility.

2. Principles of Extraction

Since most continuous extraction methods use countercurrent contacts between two phases, one a light liquid, the other a heavy one, many of the fundamentals of countercurrent gas absorption and distillation carry over to extraction. But the equilibrium relationships in liquid extraction are more difficult.

a. Ponchon–Savarit Method

Since the graphical mixing construction and the lever rule both apply to triangular diagrams, the complete Ponchon–Savarit method carries over unchanged to these coordinates. Typical constructions for type I and type II systems are shown in Figures M10 and M11, respectively. Points P and J have the same significance as in leaching. With triangular coordinates, remember that stream magnitudes and concentrations are based upon total streams.

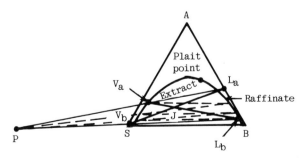

FIGURE M10. Ponchon–Savarit for type I extraction.

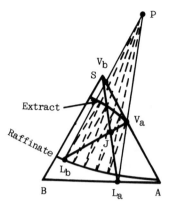

FIGURE M11.   Ponchon–Savarit for type II extraction.

b.   Rectangular Diagram Method for Extraction

Given a triangular diagram as in Figure M12a; we replot it as in Figure M12b, where F and O are the two components to be separated, and S is the solvent. Using Figure M12b, we obtain the rectangular diagram of Figure M13. Notice that all points on the x axis are solvent free (the $X_O = 0\%$ is the same as

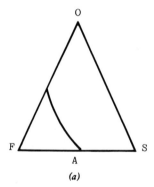

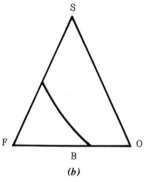

FIGURE M12.   Extraction diagrams: (a) triangular; (b) replotted triangular.

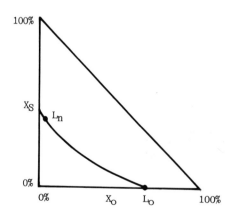

FIGURE M13.   Extraction diagram: rectangular.

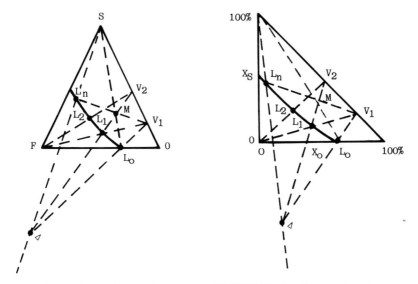

FIGURE M14a. Construction of stages: triangular diagram.

FIGURE M14b. Construction of stages: rectangular diagram.

100%F). The line from $100\% \, X_S$ to $100\% \, X_O$ is the line of 0%F. If the curved line L_0–L_n is the locus of raffinate, we know that, if F is not soluble in S, the locus of extracts will be on the S–O straight line on the triangular diagram (or on the $100\% \, X_S$–$100\%X_O$ straight line on the rectangular diagram).

c. Countercurrent Construction of Stages

The construction of stages is the same whether you use triangular coordinates, as in Figure M14a, or the rectangular coordinates of Figure M14b.

d. Minimum Solvent Ratio

(For processes with a component that is insoluble in the solvent.) Given a countercurrent process such as that shown in Figure M15, a total balance around the system gives

$$L_0 + V_{n+1} = V_1 + L_n$$

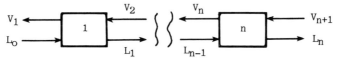

FIGURE M15. Staged countercurrent extraction system.

and if one of the components of the L stream is not soluble in the solvent, we may write a balance on that component as (call it a)

$$L_0 x_a^0 = L_n x_a^n; \qquad x_a^0 = \frac{\text{fraction of A}}{\text{in stream 0}}$$

We write a solvent balance

$$L_0 x_s^0 + V_{n+1} x_s^{n+1} = V_1 x_s^1 + L_n x_s^n$$

Now if there is no solvent in the feed, $x_s^0 = 0$, and if the solvent is pure solvent $x_s^{n+1} = 1$, so the solvent balance becomes

$$V_{n+1} = V_1 x_s^1 + L_n x_s^n$$

Define the solvent to feed ratio as P, so $P = V_{n+1}/L_0$. Then the total material balance becomes

$$1 + P = (V_1/L_0) + (L_n/L_0) = \bar{V}_1 + \bar{L}_n$$

and the solvent balance

$$P = \bar{V}_1 x_s^1 + \bar{L}_n x_s^n$$

Solving the total balance for $\bar{V}_1$, plugging it into the solvent balance, and re-arranging, gives

$$P = (1 - \bar{L}_n) \frac{x_s^1}{1 - x_s^1} + \bar{L}_n \frac{x_s^n}{1 - x_s^1}$$

We see that if $\bar{L}_n$ and x_s^n are fixed by the process, the minimum P occurs for $x_s^1 = 0$.

IV.   GAS ABSORPTION

We have a soluble vapor mixed with an inert gas. In gas absorption, a liquid in which the solute gas is soluble, is passed countercurrent to the gas stream. This unit operation is most often carried out in a packed tower, but plate columns are sometimes used.

A.   Packed Towers

1.   Principles

a.   Material Balances

In a packed tower there are no discrete sudden composition changes as there are in a plate column. The composition changes are continuous. Nevertheless, a

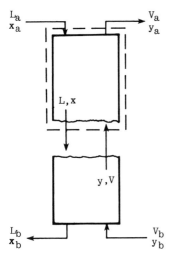

$$L_a \qquad V_a$$
$$x_a \qquad y_a$$

$$L, x$$

$$y, V$$

$$L_b \qquad V_b$$
$$x_b \qquad y_b$$ **FIGURE M16.** Packed absorption column.

total material balance is the same as for a plate column. A total balance around part of the packed column of Figure M16 gives

$$L_a + V = L + V_a$$

Likewise, a soluble component A balance is

$$L_a x_a + Vy = Lx + V_a y_a$$

with V the molal flow rate of the gas phase and L the liquid phase rate. V, L, y, and x are all at the same unspecified location.

We can also write the overall balances

$$L_a + V_b = L_b + V_a$$

$$L_a x_a + V_b y_b = L_b x_b + V_a y_a$$

Based upon these balances, the operating line may be written as

$$y = \frac{L}{V} x + \frac{V_a y_a - L_a x_a}{V}$$

or

$$L'\left(\frac{x_a}{1 - x_a} - \frac{x}{1 - x}\right) = V'\left(\frac{y_a}{1 - y_a} - \frac{y}{1 - y}\right)$$

on a solute-free basis, where

$$L' = L(1 - x), \qquad V' = V(1 - y)$$

b. Limiting Gas–Liquid Ratio

If we reduce the liquid flow at constant gas rate, the operating line slope L/V decreases as shown in Figure M17. The operating line top then moves in the

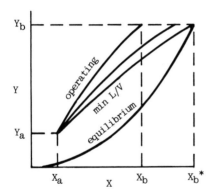

FIGURE M17. Determining limiting gas–liquid ratio.

direction of the equilibrium line. Thus the maximum liquid outlet composition and the minimum liquid rate occurs when the operating line just touches the equilibrium line at x_b^*, y_b.

c. Differential Material Balance

If we consider a thin slice of the packed column and do a differential balance on the solute, component A:

$$d(Lx) = d(Vy)$$

Note that each of these terms is the rate at which A is transferred from one phase to the other through the interface between the phases. So we may write

$$d(Lx) = d(Vy) = dN_a$$

where N_a is the transfer rate of component A in moles per hour.

2. Absorption Rate

a. Theory of Double Resistance

Solute A must go in series through a gas resistance, the interface, and a liquid resistance. If the interface offers no resistance, the gas and liquid phases are at equilibrium there. This is usually true.

Look at a slice of a packed absorption column dz at a distance z from the top. The rate of absorption is dN_a and the area of the interface is dA. Then, assuming equimolal diffusion of one component,

$$dN_a = \frac{k_y(y - y_i)}{(1 - y)_L} dA$$

where k_y = gas-phase mass transfer coefficient, mol/(area)(hr)(mol fraction difference)

y, y_i = mole fraction A in bulk and interface gas phase, respectively

$$(1 - y)_L = \frac{(1 - y) - (1 - y_i)}{\ln\left[(1 - y)/(1 - y_i)\right]}$$

Considering the liquid phase, we can also write

$$dN_a = k_x(x_i - x)\, dA$$

with

$$k_x = \text{liquid-phase mass transfer coefficient}$$

$$\left(\frac{\text{mol}}{\text{area hr mol fraction difference}}\right)$$

Thus we may write

$$dN_a = \frac{k_y(y - y_i)}{(1 - y)_L}\, dA = k_x(x_i - x)\, dA = d(Vy) = d(Lx)$$

and we may choose sets of two at will. For instance, the gas-phase equation is

$$k_y \frac{(y - y_i)}{(1 - y)_L}\, dA = d(Vy)$$

and the liquid-phase equation is

$$k_x(x_i - x)\, dA = d(Lx)$$

We define $dA = aS\, dz$ where a is the interface area per unit volume of packing, S is the cross-sectional area of the tower, and dz is the length increment.
In addition, since $V = V'/(1 - y)$,

$$d(Vy) = V'd\left(\frac{y}{1 - y}\right) = V' \frac{dy}{(1 - y)^2} = V \frac{dy}{1 - y}$$

we have

$$\frac{k_y a(y - y_i)}{(1 - y)_L}\, dz = G_{My} \frac{dy}{1 - y}$$

with $G_{My} = V/S = $ molal mass velocity of gas, and, for the liquid-phase equation

$$k_x a(x_i - x)\, dz = G_{Mx} \frac{dx}{1 - x}$$

with $G_{Mx} = $ molal mass velocity of liquid

b.   General Case—$\Delta x\, \Delta y$ Triangle

Since the transfer rate is the same for both phases,

$$G_{My} \frac{dy}{1 - y} = G_{Mx} \frac{dx}{1 - x}$$

it follows that

$$\frac{y - y_i}{x_i - x} = \frac{(1 - y)_L k_x a}{k_y a}$$

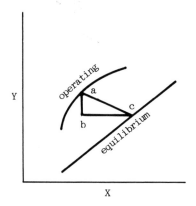

Y

X

FIGURE M18. Packed column—$\Delta x \Delta y$ triangle method.

or

$$y = -\frac{k_x a(1-y)_L}{k_y a}x + y_i + \frac{k_x a(1-y)_L}{k_y a}x_i$$

This equation has a slope of

$$-\frac{k_x a(1-y)_L}{k_y a}$$

and goes through the point (y_i, x_i) on the equilibrium curve (point c of Figure M18). So ab is $y - y_i$ and bc is $x_i - x$. By drawing many of these triangles we know $y - y_i$ and $x_i - x$ as a function of x or y. We use these data to integrate our equations graphically:

$$\int_{y_a}^{y_b} \frac{(1-y)_L\, dy}{(1-y)(y-y_i)} = \frac{k_y a}{G_{My}} \int_0^{Z_T} dz = \left(\frac{k_y a}{G_{My}}\right) Z_T$$

and

$$\int_{x_a}^{x_b} \frac{dx}{(1-x)(x_i-x)} = \left(\frac{k_x a}{G_{Mx}}\right) Z_T$$

Both equations give the same result.

c. Overall Coefficients

The general method works for either curved or straight equilibrium lines. But it uses the individual coefficients that are difficult to determine experimentally. If the equilibrium line is straight, easily obtained overall coefficients may be used.

The overall coefficients are defined by

$$K_y = \frac{dN_a/dA}{y - y^*}$$

for the overall gas-resistance coefficient, and

$$K_x = \frac{dN_a/dA}{x^* - x}$$

for the overall liquid-resistance coefficient. The overall coefficients are related to the individual coefficients by

$$\frac{1}{K_y a} = \frac{1}{k_y a} + \frac{m}{k_x a}$$

and

$$\frac{1}{K_x a} = \frac{1}{k_x a} + \frac{1}{mk_y a} = \frac{1}{mK_y a}$$

where $m = (y_i - y^*)/(x_i - x)$ is the slope of the equilibrium line. In terms of the overall coefficients,

$$\int_{y_a}^{y_b} \frac{dy}{(1 - y)(y - y^*)} = \left(\frac{K_y a}{G_{My}}\right) Z_T$$

and

$$\int_{x_b}^{x_b} \frac{dx}{(1 - x)(x^* - x)} = \left(\frac{K_x a}{G_{Mx}}\right) Z_T$$

d. Height of Transfer Unit (HTU) Method

With this method, the total height of packing Z_T is

$$Z_T = N_t H$$

with N_t the number of transfer units and $H = $ HTU. The number of transfer units may be defined by

$$N_{ty} = \int_{y_a}^{y_b} \frac{dy}{(1 - y)(y - y_i)}$$

or

$$N_{toy} = \int_{y_a}^{y_b} \frac{dy}{(1 - y)(y - y^*)}$$

or

$$N_{tx} = \int_{x_a}^{x_b} \frac{dx}{(1 - x)(x_i - x)}$$

or

$$N_{tox} = \int_{x_a}^{x_b} \frac{dx}{(1 - x)(x^* - x)}$$

Likewise we have the following equations for the HTU:

$$H_y = G_{My}/k_y a$$

$$H_x = G_{Mx}/k_x a$$

$$H_{oy} = G_{My}/K_y a$$

$$H_{ox} = G_{Mx}/K_x a$$

The overall HTU's are related to the individual HTU's by

$$H_{ox} = H_x + \left(\frac{G_{Mx}}{mG_{My}}\right)H_y = \left(\frac{G_{Mx}}{mG_{My}}\right)H_{oy}$$

e.   Controlling Resistance

With a slightly soluble gas, the slope of the equilibrium line m is large, and H_{ox} is close to H_x if G_{Mx}/G_{My} is not large. Conversely, for soluble gases, the m is small and H_{oy} is nearly equal to H_y if G_{My}/G_{Mx} is not large.

f.   Lean Gases

For systems with a small y, we have a lean gas. For these systems, the operating line is essentially straight. If, in addition, we have a straight equilibrium line, all of our previous equations (and integrations) are simpler. For this case, we have

$$G_{My}(y_b - y_a) = K_y a(y - y^*)_L Z_T$$

and

$$G_{Mx}(x_b - x_a) = K_x a(x^* - x)_L Z_T$$

where

$$(y - y^*)_L = \frac{(y_b - y_b^*) - (y_a - y_a^*)}{\ln\left[(y_b - y_b^*)/(y_a - y_a^*)\right]}$$

and

$$(x^* - x^*)_L = \frac{(x_b^* - x_b) - (x_a^* - x_a)}{\ln\left[(x_b^* - x_b)/(x_a^* - x_a)\right]}$$

In addition, the HTU equations become

$$N_{toy} = (y_b - y_a)/(y - y^*)_L$$

and

$$N_{tox} = (x_b - x_a)/(x^* - x)_L$$

3.   Coefficients and HTU's

Little general data exist.

B. Plate Columns

Gas absorption in a staged system is straightforward. For a straight operating line and a straight equilibrium line, the absorption factor (Kremser) equation is applicable. If either, or both, lines are curved, the McCabe-Thiele graphical method is appropriate.

V. HUMIDIFICATION PROCESSES

A. Definitions

The *humidity* $\mathcal{H}$ is defined to be the mass of vapor carried by each unit of mass of *vapor-free* gas, or

$$\mathcal{H} = \frac{M_a \bar{p}_a}{M_b(p_t - \bar{p}_a)}$$

where M_a, M_b are component molecular weights (A being the vapor), $\bar{p}_a$ is the partial pressure of A, and p_t is the total pressure. We may use this equation with Dalton's law to find the mole fraction y of the vapor in the gas phase

$$y = \frac{\mathcal{H}/M_a}{(1/M_b) + (\mathcal{H}/M_a)}$$

When the vapor is in equilibrium with the liquid at the gas temperature we have a *saturated gas*. According to Dalton's law the partial pressure of the vapor in the saturated gas is equal to the vapor pressure of the liquid at the gas temperature. Thus

$$\mathcal{H}_s = \frac{M_a p_a}{M_b(p_t - p_a)}$$

where p_a = vapor pressure of the liquid.

The *relative humidity* $\mathcal{H}_r$ is defined to be the ratio of the partial pressure of the vapor to the vapor pressure of the liquid, expressed as a percentage, or

$$\mathcal{H}_r = \left(\frac{\bar{p}_a}{p_a}\right) 100$$

On the other hand, the *percentage humidity* (sometimes called absolute humidity) $\mathcal{H}_a$ is defined to be the ratio of the actual humidity to the saturation humidity, expressed as a percentage,

$$\mathcal{H}_a = (100)\frac{\mathcal{H}}{\mathcal{H}_s} = 100\left(\frac{\bar{p}_a}{p_t - \bar{p}_a}\right)\left(\frac{p_t - p_a}{p_a}\right) = \mathcal{H}_r \frac{p_t - p_a}{p_t - \bar{p}_a}$$

The *humid heat* C_s is the heat necessary to raise the temperature of one pound of dry gas and the accompanying vapor by $1°F$, or

$$C_s = C_{pb} + C_{pa}\mathscr{H}$$

. where C_{pa}, C_{pb} are the specific heats of vapor and gas.

The *dew point* is the temperature to which a vapor–gas mixture must be cooled (at constant humidity) to become saturated. Note that the dew point of a saturated gas phase is the same as the gas temperature.

The *humid enthalpy* or *total enthalpy* H_y is the enthalpy of one pound of gas plus the vapor it contains, or

$$H_y = C_s(T - T_0) + \mathscr{H}\lambda_0$$

where λ_0 is the latent heat of the liquid at T_0, a datum temperature.

The adiabatic-saturation temperature T_s is the final steady-state liquid temperature for a process where a gas, initially at a humidity $\mathscr{H}$ and temperature T, flows through a liquid spray chamber where the gas is cooled and humidified adiabatically. An energy balance for the process gives the equation

$$\frac{\mathscr{H} - \mathscr{H}_s}{T - T_s} = -\frac{C_s}{\lambda_s} = -\frac{C_{pb} + C_{pa}\mathscr{H}}{\lambda_s}$$

with C_s and λ_s, evaluated at T_s.

B.   The Humidity Chart

For standard pressure, humidity charts may be used. Perry has charts for air--water as well as for several other systems.

C.   Theory of Wet-Bulb Temperature

At the wet-bulb temperature, the rate of heat transfer from the gas to the liquid is set equal to the rate of vaporization times the sum of the vapor sensible heat and the latent heat of vaporization

$$q = M_a N_a \{\lambda_w + C_{pa}(T - T_w)\}$$

where q is the sensible heat transfer rate to the liquid, the molal rate of vaporization is N_a, and λ_w is the liquid's latent heat at the wet-bulb temperature T_w.

We can also write this heat transfer rate in the usual form

$$q = h_y A(T - T_i)$$

where h_y is the gas-to-liquid surface heat transfer coefficient, the interface temperature is T_i, and A is the surface area of the liquid.

The mass transfer rate, as for gas absorption, can be written

$$N_a = k_y A \frac{(y_i - y)}{(1 - y)_L}$$

with the y's as mole fraction of vapor. Expressing the mole fractions in terms of humidities while assuming that the interface temperature is the same as the wet-bulb temperature,

$$h_y(T - T_w) = \frac{k_y}{(1 - y)_L}\left(\frac{\mathscr{H}_w}{(1/M_b) + (\mathscr{H}_w/M_a)}\right.$$
$$\left. - \frac{\mathscr{H}}{(1/M_b) + (\mathscr{H}/M_a)}\right)[\lambda_w + C_{pa}(T - T_w)]$$

With the reasonable simplifying assumptions that $(1 - y)_L \cong 1$, $C_{pa}(T - T_w) \ll \lambda_w$, and that $\mathscr{H}_w/M_a$ and $\mathscr{H}/M_a \ll 1/M_b$,

$$h_y(T - T_w) = M_b k_y \lambda_w(\mathscr{H}_w - \mathscr{H})$$

or

$$\frac{\mathscr{H} - \mathscr{H}_w}{T - T_w} = -\frac{h_y}{M_b k_y \lambda_w}$$

and for air–water systems, the Lewis relation

$$h_y/M_b k_y \cong C_s$$

is valid.

D. Humidification Processes

1. Introduction

The various types of humidification processes are shown on the humidity chart of Figure M19. Heat flow and diffusion of vapor through the gas interface characterize the wet-bulb temperature process. If the liquid is a constant temperature, the adiabatic humidifier is a wet-bulb temperature process. But with dehumidifiers and liquid coolers, if the temperature of the liquid is changing, there will also be a heat flow in the liquid phase. With heat flow a temperature gradient results, and thus a liquid phase resistance to heat flow.

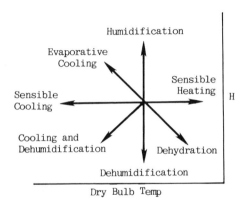

FIGURE M19. Humidification processes.

2. Gas–Liquid Interaction

The more important humidification processes can be characterized by

	Adiabatic humidifier	$T_y > T_i;$	$T_x < T_i;$	$\mathcal{H}_i > \mathcal{H}$
	Dehumidifier	$T_y > T_i;$	$T_x < T_i;$	$\mathcal{H} > \mathcal{H}_i$
	Countercurrent cooling tower			
	Upper part	$T_y < T_i;$	$T_x > T_i;$	$\mathcal{H}_i > \mathcal{H}$
	Lower part	$T_y > T_i;$	$T_x > T_i;$	$\mathcal{H}_i > \mathcal{H}$

where T_x = bulk stream liquid temperature,
   T_y = bulk stream gas temperature,
   T_i = interface temperature.
   $\mathcal{H}_i$ = interface humidity, and
   $\mathcal{H}$ = bulk gas stream humidity.

3. Packed Gas–Liquid Contactors

Consider the packed gas–liquid contactor of Figure M20. The mass velocity of the gas is G'_y, mass of vapor-free gas per cross-sectional area per hour. Let the cross section of the tower be S, so the volume of the test section is $S\, dz$. Assume the liquid is warmer than the gas. We write the following balances over the volume $S\, dz$.

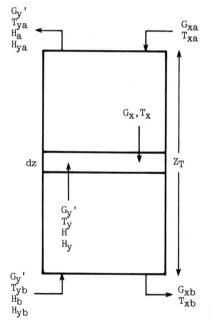

FIGURE M20. Packed gas–liquid contactor.

Enthalpy balance

$$G'_y \, dH_y = d(G_x H_x) \tag{a}$$

Heat transfer rate: liquid to interface

$$d(G_x H_x) = h_x(T_x - T_i)a_H \, dz \tag{b}$$

a_H = heat transfer area per unit volume

Heat transfer rate: interface to gas

$$G'_y C_s \, dT_y = h_y(T_i - T_y)a_H \, dz \tag{c}$$

Vapor mass transfer rate: interface to gas

$$G'_y \, d\mathscr{H} = k_y M_b(\mathscr{H}_i - \mathscr{H})a_M \, dz \tag{d}$$

a_M = mass transfer area per unit volume

With these equations and the simplifying assumptions

- G_x changes little with z
- $H_x = C_L(T_x - T_0)$; T_0 = base temperature
- $a_H = a_M = a$
- C_s changes little with $\mathscr{H}$

then

$$d(G_x H_x) = G_x \, dH_x = G_x C_L \, dT_x \tag{e}$$

which is used with Eq. (b) to give

$$G_x C_L \, dT_x = h_x(T_x - T_i)a_H \, dz$$

or

$$\frac{dT_x}{T_x - T_i} = \frac{h_x a_H}{G_x C_L} \, dz \tag{f}$$

Rearrange Eq. (c)

$$\frac{dT_y}{T_i - T_y} = \frac{h_y a_H}{C_s G'_y} \, dz \tag{g}$$

and Eq. (d)

$$\frac{dH}{H_i - H} = \frac{k_y M_b a_M}{G'_y} \, dz \tag{h}$$

Using Eq. (a) with Eq. (e)

$$\frac{dH_y}{dT_x} = \frac{G_x C_L}{G'_y}$$

Now we multiply Eq. (d) by λ_0 and then add Eq. (c) to get

$$G_y'(C_s\,dT_y + \lambda_0\,d\mathcal{H}) = (\lambda_0 k_y M_b(\mathcal{H}_i - \mathcal{H})a_M + h_y(T_i - T_y)a_H)\,dz \qquad \text{(i)}$$

Substituting

$$dH_y = C_s\,dT_y + \lambda_0\,d\mathcal{H}$$

and using the simplifying assumptions, Eq. (e) gives

$$G_y'\,dH_y = [\lambda_0 k_y M_b(\mathcal{H}_i - \mathcal{H})a + h_y(T_i - T_y)a]\,dz \qquad \text{(j)}$$

If we now direct our attention to the usual air–water system, using the Lewis relation to eliminate h_y, Eq. (j) becomes

$$G_y'\,dH_y = k_y M_b a[(\lambda_0 \mathcal{H}_i + C_s T_i) - (\lambda_0 \mathcal{H} + C_s T_y)]\,dz \qquad \text{(k)}$$

Using

$$H_i = \lambda_0 \mathcal{H}_i + C_s(T_i - T_0)$$

where H_i gives the air enthalpy at the interface, and the prior equation for H_y, Eq. (k) becomes

$$G_y'\,dH_y = k_y M_b a(H_i - H_y)\,dz$$

or finally

$$\frac{dH_y}{H_i - H_y} = \frac{k_y M_b a}{G_y'}\,dz \qquad \text{(l)}$$

Now, from Eqs. (a) and (b)

$$G_y'\,dH_y = h_x(T_x - T_i)a\,dz$$

Solving this for dz and substituting into Eq. (l) gives

$$\frac{H_i - H_y}{T_i - T_x} = -\frac{h_x}{k_y M_b} = -\frac{h_x C_s}{h_y} \qquad \text{(m)}$$

and we have used the Lewis relation again.

Finally, dividing Eq. (l) by Eq. (g)

$$\left(\frac{dT_y}{T_i - T_y}\right) \Big/ \left(\frac{dH_y}{H_i - H_y}\right) = \frac{h_y}{k_y M_b C_s}$$

But, the Lewis relation defines the right-hand side to be unity, thus

$$\frac{dT_y}{dH_y} = \frac{T_i - T_y}{H_i - H_y} \qquad \text{(n)}$$

Our working equations for the air–water system are Eqs. (l), (m), and (n).

4. Adiabatic Humidification

Because the air leaving an adiabatic humidifier may not be saturated (if it were, it would be an equilibrium problem), we must use rate equations to determine the size of a packed tower. To develop the working equations for this system, we assume:

- inlet and outlet water temperatures are equal,
- makeup water enters at the adiabatic-saturation temperature,
- $a_M = a_H = a$,
- wet-bulb and adiabatic-saturation temperatures are constant and equal and
- $T_{xa} = T_{xb} = T_i = T_x = T_s = $ constant

with $T_s = $ inlet air adiabatic-saturation temperature.
  With these assumptions, Eq. (g) becomes

$$\int_{T_{yb}}^{T_{ya}} \frac{dT_y}{T_s - T_y} = \int_0^{Z_T} \frac{h_y a}{C_s G_y'} dz$$

which integrates to

$$\ln\left(\frac{T_{yb} - T_s}{T_{ya} - T_s}\right) = \frac{h_y a Z_T}{C_s G_y} = \frac{h_y a S Z_T}{C_s G_y' S} = \frac{h_y a V_T}{C_s \dot{m}'}$$

with $V_T = S Z_T = $ contact volume and
    $\dot{m}' = G_y' S = $ total dry air flow.

Had we used the mass transfer equation (h),

$$\frac{d\mathcal{H}}{\mathcal{H}_s - \mathcal{H}} = \frac{k_y M_b a}{G_y'} dz$$

which can be integrated since $\mathcal{H}_s$, the saturation humidity at T_s, is constant:

$$\ln \frac{\mathcal{H}_s - \mathcal{H}_b}{\mathcal{H}_s - \mathcal{H}_a} = \frac{k_y M_b a}{G_y'} Z_T = \frac{k_y M_b a V_T}{\dot{m}'}$$

Notice that we may cast these equations into the HTU form, since, by definition

$$N_t = \ln \frac{\mathcal{H}_s - \mathcal{H}_b}{\mathcal{H}_s - \mathcal{H}_a}$$

$$= \text{number of humidity transfer units.}$$

Then, by the definition of H_t,

$$H_t = \frac{Z_T}{N_t} = \frac{G_y'}{k_y M_b a}$$

$$= \text{height of one humidity transfer unit}$$

Of course, we get the same answer if we base the number of transfer units on heat transfer

$$N_t = \ln\left(\frac{T_{yb} - T_s}{T_{ya} - T_s}\right), \qquad H_t = \frac{G_y' C_s}{h_y a}$$

VI. DRYING OF SOLIDS

A. Introduction

Many types of equipment are used for drying a wide variety of solids. Thus there is not a single theory of drying. Most dryers are designed using general principles with experimental data.

B. Equilibria

The moisture content of a solid is expressed as the mass of water per unit mass of bone-dry solid. So the driving force for mass transfer in a wet solid is the free-moisture content X, which is the difference between the total moisture content X_T and the equilibrium moisture content X^*.

C. Rate of Drying

The capacity of a dryer depends upon both the rate of mass transfer and the rate of heat transfer. The heat of vaporization must be supplied to the vaporization zone to vaporize the water. The water may be either at or near the solid surface or within the solid, depending upon the solid and the conditions.

D. Rate-of-Drying Curves

With constant dryer conditions, the moisture content X_T of a solid usually decreases, as shown in Figure M21. The graph is linear at first, then curves up, and finally levels off. Figure M22 shows the drying rate. It is horizontal at first, then it curves downward and eventually reaches zero. Combining the two figures, we obtain, for different materials, the drying-design curves of Figure M23.

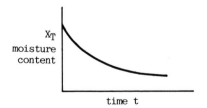

FIGURE M21. Moisture-content curve.

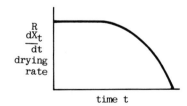

FIGURE M22. Drying-rate curve.

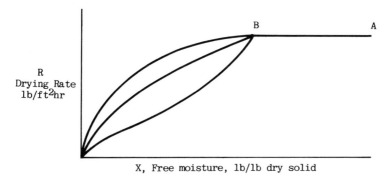

FIGURE M23. Drying-design curve.

The differences in the shapes are the result of differences in the mechanism of moisture flow in the different materials.

1. Constant-Rate Period

Every drying-rate curve has at least two distinct parts. After a warm-up period (not shown in the figure), each curve has the horizontal line AB. This period is the constant-rate period. The rate of drying is independent of moisture content. The solid is so wet that a continuous film of water covers the entire drying surface and this water behaves as if the solid were not there. If there is no heat transfer by radiation or by direct contact of the solid with hot surfaces, the temperature of the solid during the constant-rate period is the wet-bulb temperature of the drying air.

2. Critical Moisture Content and Falling-Rate Period

As the moisture content decreases, the constant-rate period ends at a definite moisture content. With further drying, the rate decreases. The point at the end of the constant-rate period (point B) is called the critical point. At this point, there is not enough liquid water on the surface to maintain a continuous film over the solid. Of course, if the initial moisture content of the solid is below the critical, there will be no constant-rate period.

The period after the critical point is called the falling-rate period. The shape of this curve varies from one type of solid to another. If also depends upon processing conditions and on the solid thickness.

E. Calculation of Drying Time

The size of the equipment needed for a given capacity is fixed by the drying time. With constant drying conditions, the drying time can be determined from the rate-of-drying curve. If we consider the drying of a slab of solid of half-thickness

s and face area S dried from both sides, the total area for drying is $2S$. Now the ordinate R of the rate-of-drying curve is

$$R = -\frac{dm}{2S\,dt}$$

where m = mass of total moisture in the solid and
t = drying time.

The abscissa of the curve is X, the mass of free moisture per unit mass of bone-dry solid. If ρ_s is the density of the solid, in mass per unit volume of bone-dry material, assumed constant, then

$$dm = 2sS\rho_s\,dX$$

Thus

$$R = -s\rho_s\frac{dX}{dt}$$

Integrating this equation between X_1 and X_2, the initial and final free-moisture contents,

$$t_T = s\rho_s \int_{X_2}^{X_1} \frac{dX}{R}$$

where t_T is the total drying time. The integration may be performed numerically using the rate-of-drying curve. It can be done analytically if R is known as a function of X.

1. Equations for Constant-Drying Conditions

The equation may be formally integrated for two conditions: during the constant-rate period and during the falling-rate period, if the rate-of-drying curve is linear.

During the constant-rate period, R is constant at its value R_c. Therefore, the time of drying in the constant-rate period t_c is given by

$$t_c = s\rho_s \int_{X_2}^{X_1} \frac{dX}{R_c} = s\rho_s \frac{(X_1 - X_2)}{R_c}$$

When the drying rate is linear in X during the falling-rate period

$$R = aX + b; \qquad dR = a\,dX$$

The integral becomes

$$t_f = \frac{s\rho_s}{a} \int_{R_2}^{R_1} \frac{dR}{R} = \frac{s\rho_s}{a} \ln\left(\frac{R_1}{R_2}\right)$$

Of course, the constant a is the slope of the rate-of-drying curve or

$$a = \frac{R_c - R'}{X_c - X'}$$

where R_c = rate at critical point,
R' = rate at final processing condition,
X_c = free-moisture content at critical point, and
X' = free-moisture content at final processing condition.

So

$$t_f = \frac{s\rho_s(X_c - X')}{R_c - R'} \ln\left(\frac{R_1}{R_2}\right)$$

When the drying process covers both a constant-rate period and a linear falling-rate period, the total drying time is the sum of t_c and t_f.

PROBLEMS

A. How many cubic feet per minute of entering air is needed to evaporate 10 lb of water per hour from a rayon, if the air enters at 80°F and 25% humidity and leaves at 170°F and 55% relative humidity. The operating pressure is 14.3 psia.

B. A nitrogen–hydrogen chloride mixture (10 vol% hydrogen chloride) is to be scrubbed with water to remove the hydrogen chloride. To satisfy environmental concerns, 99% of the inlet HCl must be removed. Assuming that the gas leaving the scrubber will be at 125°F and 1 atm saturated with water, what will be the volume and composition of the gas leaving if we must process 200 lb mol/hr of dry entering gas.

C. We wish to extract nicotine from water using kerosene. If we have 100 lb of a 2% nicotine solution extracted once with 200 lb of kerosene, what percentage of the nicotine will be extracted?

$$\text{Equilibrium:} \quad Y \frac{\text{lb nicotine}}{\text{lb kerosene}} = 0.90 X \frac{\text{lb nicotine}}{\text{lb water}}$$

D. We must cool 2500 gal/min (gpm) of water from 120 to 80°F. A cooling tower, at 70°F and 1 atm, is to be designed to operate with entering air of 40% relative humidity.

    (a) How many cubic feet per minute (cfm) of entering air must be supplied?

    (b) How many gallons per minute of makeup water must be supplied if windage loss is 95% of the amount lost by evaporation?

E. A solvent extraction will be performed on a solid B that contains a soluble component A (mass fractions $X_a = 0.25$, $X_b = 0.75$). The solvent to be used, C, is mutually insoluble with solid B. If we neglect the entrainment of B in the overflow solution and screw press the extracted solid to 1 lb of solution per pound of B, calculate the pounds of solute-A-free solvent C which we must feed the extractor, per pound of A–B feed, in order to obtain 95 % of the solute A in the extract overflow solution. What is the concentration of A in the resulting extract overflow solution?

F. We wish to scrub acetone from an air stream containing 0.020 mol fraction acetone. If we use a countercurrent packed scrubber designed so that the exit gas does not exceed 0.0002 mol fraction acetone,

   (a) what is the number of overall gas-phase transfer units;
   (b) what is the height of one transfer unit;
   (c) what is the total height of packing used?

Base your calculation upon the following data:
   Gas rate is 700 ft³/hr per square foot cross section at standard conditions;
   Water rate is 1000 lb/hr ft²;
   $K_g a$ is 1.75 lb mol/ft³ hr mol fraction difference;
   Henry's law is valid and $y_e = 1.75x_e$ (y_e is the acetone vapor in equilibrium with the mole fraction of acetone x_e in the liquid).

G. We wish to make our office building more comfortable by conditioning the air. We feel that the scheme shown in Figure M24 is appropriate.

Air to building	10^5 ft³/hr at 70°F, 50 % relative humidity
Recycle air	5×10^4 ft³/hr at 70°F, 50 % relative humidity
Fresh air	40°F, 10 % relative humidity
Steam	30 psig

The humidifier will be designed so that the air leaving it will be at 85 % humidity.

Design coefficients: humidifier $h_y a = 80$ Btu/ft³ hr °F
                        preheater $U = 20$ Btu/ft² hr °F

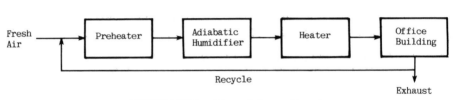

FIGURE M24.   Schematic for problem G.

Based upon this scheme, find:

(a) Cubic feet per minute of fresh air required, at entrance conditions.

(b) Temperature and absolute humidities at (i) preheater inlet, (ii) humidifier inlet, and (iii) humidifier outlet.

(c) Temperature of water required in the humidifier.

(d) Humidifier volume.

H. We need to cool 11,000 gal/min of process water from 100 to 80°F. Our engineer suggests a mechanical (induced) draft cooling tower with 2 cells, each 45 × 45 ft × 55-ft high. She wants to use air at a rate of 750,000 ft³/min for each cell (measured at 80°F dry bulb and 70°F wet bulb). The design wet-bulb temperature was 70°F. Is this design approximately correct? If $h_y a = 10$ Btu/ft³ hr °F, calculate the tower height required.

I. How many stages and how much water is needed for the counter-current extraction of NaOH from a feed consisting of 80 lb NaOH, 400 lb H_2O, and 100 lb $CaCO_3$. The final extract solution will contain 10% NaOH with the recovery of 95% of the NaOH. We shall set the underflow at a constant 3.0-lb solution/lb $CaCO_3$.

J. We have a binary mixture, 50% (mol) A, to feed to a distillation column. We need 97% A overhead and a minimum of 97% B in the bottoms. For an atmospheric system with a total condenser and a reboiler and a temperature such that the feed is half vaporized as it enters the column, determine (a) the minimum reflux ratio; (b) the theoretical stages needed at total reflux; (c) the theoretical stages needed using a reflux ratio of two times the minimum; and (d) the feed plate number.

The equilibrium data are

x_A	0	0.1	0.2	0.3	0.4	0.5	0.6	0.7
y_A	0	0.242	0.418	0.551	0.657	0.741	0.811	0.870

x_A	0.8	0.9	1.0
y_A	0.920	0.963	1.0

K. An ideal mixture of A and B is to be distilled continuously. If the relative volatility α is constant at 1.75 and

$$\text{feed} = \text{saturated liquid with 50 mol\% A,}$$
$$\text{feed rate} = 200 \text{ lb mol/hr,}$$
$$\text{distillation composition} = 90\% \text{ A,}$$
$$\text{bottoms composition} = 10\% \text{ A,}$$

find analytically the minimum number of theoretical stages and the minimum reflux ratio.

L. When a porous solid was dried under constant-drying conditions, 5 hours are required to reduce the moisture from 30 to 12 lb H_2O/lb dry solid. Critical moisture content is 18 lb H_2O/lb dry solid and the equilibrium moisture is 5 lb H_2O/lb dry solid. If the drying rate during the falling-rate period is a straight line through the origin, determine the time needed to dry the solid from 30 to 8 lb H_2O/lb dry solid.

PROBLEM-SOLVING STRATEGIES

A. Although this seems to be a drying problem, it is actually just a material balance using humidities. But be careful—both humidity and relative humidity are used. Check the definitions. Remember that most humidity charts are based on atmospheric pressure, not 14.3 psia. So you must perform calculations instead of using a chart.

B. This scrubber problem is just a material balance problem. The leaving gas, saturated with water vapor, will have 99% of the inlet HCl. A humidity chart will give you the outlet lb H_2O vapor/lb dry air. Since this is mostly nitrogen, the humidity must be calculated. At standard conditions, the volume occupied by 1 lb mol of any gas is 359 ft^3/lb mol.

C. This is an extraction problem, with one stage. We must assume that it is an ideal stage. The given equilibrium equation is a straight line. Since, on a nicotine-free basis, the material balance is also a straight line, the problem could be solved analytically using the Kremser equation (sometimes called the Souders-Brown equation). But with one stage, writing a nicotine balance and using the equilibrium equation is quicker. Since there is only one stage, this extraction problem is just a simple material balance problem.

D. This cooling tower must be designed and there is not enough information given to assume that it is simply a material balance problem. We need a design strategy. The outlet air should be saturated with water vapor. But to calculate the saturated humidity of the outlet air, we need to know the outlet air temperature in order to find the correct vapor pressure. Heat transfer ideas suggest that it is unlikely that the outlet air temperature would equal the inlet water temperature. The inlet air temperature is 10°F lower than the outlet water temperature. It is logical to assume that the top temperature approach is the same as the bottom temperature approach or 10°F. So assuming an outlet air temperature of 120°F − 10°F = 110°F fixes the humidity. A heat balance will then set the inlet flow rate of air.

E. Since this solvent extraction problem has but one stage, a material balance provides the answer. A total balance and balances on A and C make this problem a gift.

F. This is a packed scrubber, so we rule out the ideal stage concept. The equilibrium equation is a straight line, based upon mole fraction of acetone. In general, a material balance (the operating line) based upon the total moles of a stream is not a straight line. One way to get a straight-line material balance is to use acetone-free air and water streams. Should we choose this approach, the equilibrium relationship would no longer be straight. The other possibility occurs if the acetone is present in such small amounts that the material balance is approximately straight. This is the case for this problem. Because both operating line and equilibrium line are approximately straight, an analytical equation is available. This is always preferable to a numerical integration. Thus we may use the lean gas absorption equations. Material balances supply the unknown data for the equation. With the equation, the solution is simple.

G. This is a humidification problem at atmospheric pressure. A humidity chart can be used. With the data given, we know the lb H_2O/lb dry air leaving the preheater is the same as the preheater inlet since the preheater simply increases the temperature. To determine the preheater inlet humidity, we do a H_2O balance around the fresh-air–recycle-air mixing point. An energy balance sets the inlet temperature. The humidifier outlet humidity is the same as the office humidity since the heater adds no moisture. The humidifier outlet temperature is found on the chart where the humidity line intersects the 85% relative humidity line. To determine the humidifier volume, we need an adiabatic humidifier design equation. Fortunately, McCabe and Smith has one.

H. We do not have enough information given to solve this cooling tower design problem. Since this is a packed tower, we cannot use ideal stage concepts. We know that the operating line is an energy balance. We know that the equilibrium line is the enthalpy at saturation. We know that theory is available to solve for packing height, given $k_y a$. But we also know that theory is not available to determine bed cross-sectional area. We must know, or be able to find, common practice concepts for this. We must check Perry's *Chemical Engineers' Handbook*. If we find what we need, we can judge whether to attempt the problem. If not, we should move to another problem.

I. This is a staged solvent extraction problem. When looking for easy extraction problems, we want a straight material balance (operating) line and a straight equilibrium line. We remember that a constant underflow is a straight equilibrium line, and that a liquid–solid system usually gives a straight material balance line. It looks as if the Kremser equation will give us an easy solution. But we remember that with constant underflow, the inlet stream is rarely set at the same rate as the underflow. We may have to use just a material balance on the feed stage (first stage) and the Kremser equation where the underflow is indeed constant.

J. This binary distillation problem is probably best solved graphically. We could analytically find solutions for the minimum reflux and theoretical stages at total reflux if the equilibrium data were based upon a constant relative volatility α. But we would still need graphical solutions for the other answer. So we would use graphical solutions for the complete problem.

K. For the minimum number of stages in a binary distillation with constant relative volatility, we use the Fenske equation. I do not know of an analytical equation for minimum reflux ratio, but I know graphically how to obtain it. So I can develop an equation upon which the graphical solution is based. I know that the minimum reflux line is straight and that its two intersection points are (a) the distillate composition x_D on the 45° line, and (b) the point where the feed q line intersects the equilibrium line. The equilibrium line is defined by $\alpha = y(1-x)/x(1-y)$.

L. This is a drying problem. Problems like this are solved analytically or by numerical integration. Since the falling rate is a straight line, an analytical solution is possible.

SOLUTIONS

A. See Figure M25.
Note: Humidity charts are based on 1-atm pressure, so we must use equations, not charts
Vapor pressure (VP) of H_2O [McCabe and Smith, (McC & S) 3rd ed., App. 8]; at 80°F, 0.5069 psia; at 170°F, 5.992 psia. D = dryair.

$$\mathcal{H}_i = 0.25\left(\frac{0.5069}{14.3 - 0.5069}\right)\left(\frac{18}{28.9}\right) = 0.0057 \frac{\text{lb } H_2O}{\text{lb D air}}$$

$$\mathcal{H}_o = \frac{0.55(5.992)}{14.3 - 0.55(5.992)}\left(\frac{18}{28.9}\right) = 0.1865 \frac{\text{lb } H_2O}{\text{lb D air}}$$

So we remove

$$(\mathcal{H}_o - \mathcal{H}_i)\frac{\text{lb } H_2O}{\text{lb D air}} \quad \text{or} \quad 0.1808 \frac{\text{lb } H_2O}{\text{lb D air}}$$

$$\frac{\text{lb}}{\text{hr}} \text{ D air required} = \frac{10 \text{ lb } H_2O}{\text{hr } 0.1808 \text{ lb } H_2O/\text{lbD air}} = 55.3 \frac{\text{lb D air}}{\text{hr}}$$

$$\text{humid vol of inlet air} = 359\left(\frac{540}{492}\right)\left(\frac{1}{28.9} + \frac{0.0057}{18}\right)$$

$$= 13.76 \text{ ft}^3/\text{lb D air}$$

$$\text{cfm inlet air} = 55.3 \frac{\text{lb D air}}{\text{hr}} 13.76 \frac{\text{ft}^3}{\text{lb D air}} \frac{1 \text{ hr}}{60 \text{ min}}$$

$$= 12.68 \text{ cfm}$$

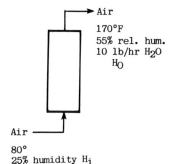

Air

170°F
55% rel. hum.
10 lb/hr H_2O
H_0

Air

80°
25% humidity H_i
no. of cfm?

FIGURE M25. Schematic for solution A.

B. See Figure M26.
   Basis: 200 lb mol/hr dry inlet gas

Gas	lb mol
HCl	20
N_2	180

Outlet gas: 180 lb mol/hr N_2; 0.01 (20) = 0.2 lb mol/hr HCl and H_2O.

$$\mathcal{H}_s = \frac{MW_{H_2O} \cdot VP_{H_2O}}{MW_{N_2} (1 \text{ atm} - VP_{H_2O})}$$

Steam tables:
$$VP_{H_2O} \text{ at } 125 = 1.942 \text{ psia}; MW_{H_2O} = 18, MW_{N_2} = 28.$$

$$\mathcal{H}_s = \frac{18}{28} \left(\frac{1.942}{14.7 - 1.942} \right) = 0.098 \frac{\text{lb } H_2O}{\text{lb D gas}}$$

or

$$\frac{1.942}{14.7 - 1.942} = 0.152 \frac{\text{mol } H_2O}{\text{mol D gas}}$$

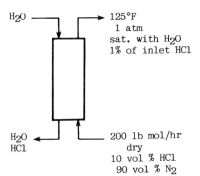

H_2O

125°F
1 atm
sat. with H_2O
1% of inlet HCl

H_2O
HCl

200 lb mol/hr
dry
10 vol % HCl
90 vol % N_2

FIGURE M26. Schematic for solution B.

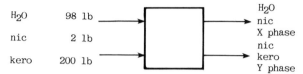

FIGURE M27. Schematic for solution C.

Outlet gas:

Gas	Moles	Percent
N_2	180	86.7
HCl	0.2	0.1
H_2O	27.43	13.2
	207.63	100.0

$$H_2O = 0.152\left(\frac{\text{mol } H_2O}{\text{mol D gas}}\right)(180 + 0.2) \text{ mol D gas}$$

$$= 27.43$$

Volume of gas leaving:

$$207.63 \frac{\text{lb mol}}{\text{hr}} \times \frac{359 \text{ ft}^3}{\text{lb mol}} \left(\frac{460 + 125}{492}\right) = 88,630 \frac{\text{ft}^3}{\text{hr}}$$

C. Nicotine balance (see Figure M27) (nic = nicotine, kero = kerosene)
input = output in X phase + output in Y phase

$$Y \frac{\text{lb nic}}{\text{lb kero}} = 0.90X \frac{\text{lb nic}}{\text{lb } H_2O}$$

$$2 \text{ lb} = 98 \text{ lb } H_2O X \frac{\text{lb nico}}{\text{lb } H_2O} + 200 \text{ lb kero } Y \frac{\text{lb nic}}{\text{lb kero}}$$

Want: pounds nicotine out.
Now $X = Y/0.9$

$$2 = 98\left(\frac{Y}{0.9}\right) + 200Y = \left(\frac{98}{0.9} + 200\right)Y;$$

$$Y = 2 \div \left(\frac{98}{0.9} + 200\right) = \frac{2}{310} = 6.45 \times 10^{-3}$$

$$\text{lb nico out} = 200 \text{ lb kero} \left(6.45 \times 10^{-3} \frac{\text{lb nic}}{\text{lb kero}}\right) = 1.29 \text{ lb}$$

so

$$\text{lb nicotine extracted} = 1.29.$$

$$\% \text{ extracted} = (1.29/2)100 = 64.5\%$$

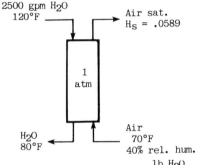

2500 gpm H₂O
120°F

Air sat.
H_S = .0589

1
atm

H₂O
80°F

Air
70°F
40% rel. hum.

$H = .0063 \dfrac{\text{lb H}_2\text{O}}{\text{lb D air}}$ **FIGURE M28.** Schematic for solution D.

D. See Figure M28.

Basis:

$$2500 \frac{\text{gal}}{\text{min}} \times 8.25 \frac{\text{lb}}{\text{gal}} = 20{,}623 \frac{\text{lb}}{\text{min}}$$

$$0.13368 \frac{\text{ft}^3}{\text{gal}} \times 61.71 \frac{\text{lb}}{\text{ft}^3} = 8.25 \frac{\text{lb}}{\text{gal}}$$

(a) Entering air. Assume same outlet approach as inlet approach 120 − $T_{ao} = 80 − 70$. So $T_{ao} = 110°F$; $\mathscr{H}_s = 0.0589$ lb H₂O/lb D air; and

$$\text{H}_2\text{O picked up/lb D air} = 0.0589 − 0.0063 = 0.0526 \frac{\text{lb H}_2\text{O}}{\text{lb D air}}$$

$$\text{heat removed from H}_2\text{O} = 20{,}623 \frac{\text{lb}}{\text{min}} \times \frac{1 \text{ Btu}}{\text{lb}°F} (120\text{-}80°F)$$

$$= 8.25 \times 10^5 \frac{\text{Btu}}{\text{min}}$$

$$C_s \text{ (inlet air)} = 0.242 \frac{\text{Btu}}{°F \text{ lb D air}}; \quad C_s \text{ (outlet air)} = 0.267 \frac{\text{Btu}}{°F \text{ lb D air}}$$

$$H = C_s(T − T_0) + \mathscr{H}\lambda_0$$

If $T_0 = 32°F$; $\lambda_0 = 1075.8$ Btu/lb,

$$H_{out} = 0.267(110 − 32) + 0.0589(1075.8) = 84.2 \text{ Btu/lb D air}$$

$$H_{in} = 0242(70 − 32) + 0.0063(1075.8) = 16.0 \text{ Btu/lb D air}$$

Heat Balance: heat removed from H₂O = heat gained by air

$$8.25 \times 10^5 \frac{\text{Btu}}{\text{min}} = (84.2 − 16.0)\left(\frac{\text{Btu}}{\text{lb D air}}\right)\left(\frac{D \text{ lb D cair}}{\text{min}}\right)$$

$$D = 1.21 \times 10^4 \frac{\text{lb D air}}{\text{min}}$$

Specific vol of inlet air = 13.3 ft^3/lb D air at 70°F, so

$$\text{cfm entering} = 1.21 \times 10^4 \frac{\text{lb D air}}{\text{min}} \frac{13.3 \text{ ft}^3}{\text{lb D air}} = 1.61 \times 10^5 \frac{\text{ft}^3}{\text{min}}$$

(b) Makeup water

$$\text{H}_2\text{O lost by evap} = 0.0526 \frac{\text{lb H}_2\text{O}}{\text{lb D air}} 1.21 \times 10^4 \frac{\text{lb D air}}{\text{min}} = 636 \frac{\text{lb H}_2\text{O}}{\text{min}}$$

$$= \frac{636 \text{ lb H}_2\text{O/min}}{8.25} \frac{}{\text{lb/gal}} = 77.2 \frac{\text{gal}}{\text{min}} = \text{evap}$$

$$\text{makeup required} = 95\% \text{ evap} + \text{evap} = 1.95 \left(77.2 \frac{\text{gal}}{\text{min}} \right)$$

$$= 150 \text{ gpm}$$

E. See Figure M29.

Basis: 1 lb of A–B feed

E is solvent and A, with no solid; U is solution (A + C) and solid; Solution is same composition in both E and U. With our basis, U is 0.75 lb solid and also 0.75 lb solution

$$\begin{aligned}
A \text{ balance:} & \quad 0.25 \text{ lb} = e_a E + u_a 0.75 \\
C \text{ balance:} & \quad S \text{ lb} = e_s E + u_s 0.75 \\
\text{total balance:} & \quad S + 0.25 = E + 0.75 \text{ or } S = E + 0.5
\end{aligned}$$

Now

$$e_a E = 0.95(0.25); \quad u_a U = 0.05(0.25); \quad u_a = \frac{0.05(0.25)}{0.75} = 0.0167$$

and

$$e_a = u_a, e_s = u_s = 1 - u_a = 1 - \frac{0.05(0.25)}{0.75} = 1 - \frac{0.05}{3} = \frac{2.95}{3} = 0.9833$$

and then

$$E = \frac{0.95(0.25)}{0.0167} = 14.25$$

Using total balance, S = 14.25 + 0.5 = 19.75.

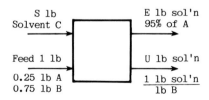

FIGURE M29.   Schematic for solution E.

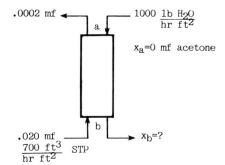

.0002 mf

1000 $\frac{\text{lb H}_2\text{O}}{\text{hr ft}^2}$

$x_a = 0$ mf acetone

.020 mf
$\frac{700 \text{ ft}^3}{\text{hr ft}^2}$ STP

$x_b = ?$

FIGURE M30. Schematic for solution F.

F. See Figure M30. This is a lean gas; we use the lean gas absorption equations

$$N_{toy} = \frac{y_b - y_a}{(y - y^*)_{LM}}, \qquad H_{oy} = \frac{G_{My}}{K_y a}, \qquad Z_T = N_{toy} \cdot H_{oy}$$

$$(y - y^*)_{LM} = \frac{(y_b - y_b^*) - (y_a - y_a^*)}{\ln\left[(y_b - y_b^*)/(y_a - y_a^*)\right]}; \qquad LM = \text{log mean}$$

$$y_b^* = 1.75x_b, \qquad y_a^* = 1.75x_a$$

$$G = \frac{700 \text{ ft}^3}{\text{hr ft}^3}\left(\frac{\text{lb mol}}{359 \text{ ft}^3}\right) = 1.950 \frac{\text{lb mol}}{\text{hr ft}^2};$$

$$L = \frac{1000 \text{ lb H}_2\text{O}}{\text{hr ft}^2}\left(\frac{\text{lb mol}}{18 \text{ lb}}\right) = 55.56 \frac{\text{lb mol}}{\text{hr ft}^2}$$

To get x_b—acetone balance

$$G(0.020 - 0.0002) = Lx_b, \qquad x_b = (1.98 \times 10^{-2}G)/L = 6.949 \text{ } 10^{-4} \text{ mf}$$

$$y_b - y_b^* = 0.020 - 1.75(6.949 \times 10^{-4}) = 0.01878$$

$$y_a - y_a^* = 0.0002 - 1.75(0) \qquad\qquad = 0.0002$$

$$(y - y^*)_{LM} = \frac{0.01878 - 0.0002}{\ln(0.01878/0.0002)} = \frac{0.01858}{4.542}$$

(a)

$$N_{toy} = \frac{(0.020 - 0.0002)}{(0.01858/4.542)} = 4.84$$

(b)

$$H_{oy} = \frac{G_{My}}{K_y a} = \frac{1.950 \text{ lb mol}}{\text{hr ft}^2} \frac{\text{ft}^3 \text{ hr}}{1.75 \text{ lb mol}} = 1.114 \text{ ft}$$

(c)

$$Z_T = 1.114 \text{ ft } (4.84) = 5.39 \text{ ft}$$

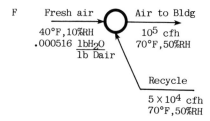

F   Fresh air   Air to Bldg

40°F,10%RH 10^5 cfh
.000516 $\frac{\text{lbH}_2\text{O}}{\text{lb Dair}}$ 70°F,50%RH

                              Recycle

                          5×10^4 cfh
                          70°F,50%RH   **FIGURE M31.** Schematic for solution G.

G. (a)   Cubic feet per minute of fresh air required—based on dry air balance
          (see Figure M31).

$$\text{humidity of air to bldg} = 0.00794 \, \frac{\text{lb H}_2\text{O}}{\text{lb D air}}$$

$$\text{humid volume} = 359\left(\frac{T}{492}\right)\left(\frac{1}{28.9} + \frac{\mathscr{H}}{18}\right) \frac{\text{ft}^3}{\text{lb D air}}$$

So

$$\text{humid volume of air to bldg} = 13.55 \, \frac{\text{ft}^3}{\text{lb D air}}$$

and

$$\text{humid volume of fresh air} = 12.63 \, \frac{\text{ft}^3}{\text{lb D air}}$$

$$\text{air to bldg} = 10^5 \, \text{cfh}/13.55 \, \frac{\text{ft}^3}{\text{lb D air}} = 7.379 \times 10^3 \, \frac{\text{lb D air}}{\text{hr}}$$

$$\text{recycle} = 0.5 \times 10^5 \, \text{cfh}/13.55 \, \frac{\text{ft}^3}{\text{lb D air}} = 3.69 \times 10^3 \, \frac{\text{lb D air}}{\text{hr}}$$

$$F + 3.690 \times 10^3 \, \frac{\text{lb D air}}{\text{hr}} = 7.379 \times 10^3 \, \frac{\text{lb D air}}{\text{hr}},$$

$$F = 3.689 \times 10^3 \, \frac{\text{lb D air}}{\text{hr}}$$

$$F = 3.689 \times 10^3 \, \frac{\text{lb D air}}{\text{hr}} \times 12.63 \, \frac{\text{ft}^3}{\text{lb D air}} \frac{1 \, \text{hr}}{60 \, \text{min}}$$

$$= 776.5 \, \text{ft}^3/\text{min inlet fresh air}$$

   (b)   Temperatures and humidities.
        (i)   Preheater inlet—composed of fresh air and recycle.

We do a mass balance for humidity and an energy balance for temperature

$$C_{PH_2Ovap} = 0.45, \qquad C_{pair} = 0.25 \text{ Btu/lb } °F \qquad \text{Mc \& S}$$

$$3.689 \times 10^3 \frac{\text{lb D air}}{\text{hr}} \left(0.000516 \frac{\text{lb H}_2\text{O}}{\text{lb D air}} \right)$$

fresh feed

$$+ 3.69 \times 10^3 (0.00794)$$

recycle

$$= 7.379 \times 10^3 \frac{\text{lb D air}}{\text{hr}} \frac{\mathcal{H} \text{ lb H}_2\text{O}}{\text{lb D air}}$$

preheater entrance

so

$$\mathcal{H} = 0.00423 \frac{\text{lb H}_2\text{O}}{\text{lb D air}}$$

$$3.689 \times 10^3 \frac{\text{lb D air}}{\text{hr}} C_s(40°F - T_0) + 3.690 \times 10^3 C_s(70 - T_0)$$

fresh air   recycle

$$= 7.379 \times 10^3 C_s(T - T_0)$$

to preheater

If all C_s are approximately equal,

$$3.689 \times 10^3(40°F) + 3.690 \times 10^3(70°F) = 7.379 \times 10^3(T)$$

or $\qquad T = 55°F$

(ii) Humidifier inlet and (iii) Humidifier outlet.
At humidifier exit, $\mathcal{H}$ is same as office bldg $\mathcal{H}$, that is, 0.00794 lb H$_2$O/lb D air and this must be at 85% relative humidity (RH) to satisfy the design. Intersection of 85% RH line and $\mathcal{H} = 0.00794$ give $T = 55°F$. At humidifier inlet, $\mathcal{H}$ is same as that for air to preheater or 0.00423. Follow adiabatic saturation line to this humidity gives $T = 70°F$.

(c) Temperature of H$_2$O required in humidifier is at 100% saturation on the adiabatic saturation or 52°F.

(d) Humidifier volume.
The adiabatic humidification design equation is given by McC & S as

$$\ln\left(\frac{T_{yb} - T_s}{T_{ya} - T_s} \right) = \frac{h_{ya}Z_T}{C_s G_y} = \frac{h_{ya}V_T}{C_s \dot{m}'}$$

where $\dot{m}'$ = total flow of dry air and
T_S = adiabatic saturation temp. of inlet air

$$T_{yb} = 70°F, \qquad T_{ya} = 55°F, \qquad T_S = 52°F$$

$$C_s(\text{at } 70°) = 0.243\,\frac{\text{Btu}}{\text{lb °F}}, \qquad C_s(\text{at } 55°) = 0.245\,\frac{\text{Btu}}{\text{lb °F}},$$

$$C_s(\text{avh}) = 0.244\,\frac{\text{Btu}}{\text{lb°F}}$$

$$\dot{m}' = 7.379 \times 10^3\,\frac{\text{lb D air}}{\text{hr}}, \qquad V_T = \frac{C_s\dot{m}'}{h_y a}\ln\left(\frac{T_{yb} - T_s}{T_{ya} - T_s}\right)$$

$$V_T = \left(\frac{0.244\ \text{Btu}}{\text{lb°F}}\right)\left(\frac{7.379 \times 10^3\ \text{lb}}{\text{hr}}\right)\left(\frac{\text{ft}^3\ \text{hr°F}}{80\ \text{Btu}}\right)\ln\left(\frac{70 - 52}{55 - 52}\right) = 40.3\ \text{ft}^3$$

H. According to Figure 15-6, p. 852 of the *first* edition of McC & S (also found in Perry's Handbook), for a hot water temperature of 100°F and a cold water temperature of 80°F, with wet-bulb temperature of 70°F, the water concentration should be 2.7 gal/min ft².

For 11,000 gpm, the cross-section area should be

$$\frac{11,000\ \text{gpm}}{2.7\ \text{gpm/ft}^2} = 4074\ \text{ft}^2$$

Our designer suggested 45 × 45 ft = 2025/cell or 4050 ft² total McC & S (same edition) suggests a superficial air velocity of about 6 ft/sec. Our designer suggested

$$\left(\frac{750,000\ \text{ft}^3}{\text{min}\ 2025\ \text{ft}^2}\right)\left(\frac{\text{min}}{60\ \text{sec}}\right) = 6.17\,\frac{\text{ft}}{\text{sec}}$$

Now, we know that we are in the ballpark with these rules of thumb. We want to find the height. McCabe and Smith gives several different (but equal) equations for doing this gas–liquid contact

$$\frac{dH_y}{H_i - H_y} = \frac{k_y a M_b}{G_y'}\,dZ$$

but for air–water, the Lewis relation gives

$$\frac{h_y a}{M_b k_y a} = C_s$$

So

$$\frac{dH_y}{H_i - H_y} = \frac{h_y a}{C_s G_y'}\,dZ$$

where H_i = saturation enthalpy (equilibrium line) and
H_y = actual air enthalpy (operating line)

or

$$Z = \frac{C_s G'_y}{h_y a} \int \frac{dH_y}{H_i - H_y}$$

If both equilibrium and operating lines were straight, we could use the enthalpy equation analogous to the *lean* gas absorption equation or

$$\int_{H_{yb}}^{H_{ya}} \frac{dH_y}{H_i - H_y} = \frac{H_{ya} - H_{yb}}{(H_{ia} - H_{ya}) - (H_{ib} - H_{yb})} \bigg/ \ln \frac{(H_{ib} - H_{yb})}{(H_{ib} - H_{yb})}$$

But we see that the H_i line, shown in Figure M32a, is not linear with temperature so we use graphical integration. The operating line is given by

$$G_y(H_y - H_{yb}) = LC_{pl}(T_x - T_{xa})$$

or

$$H_y = H_{yb} + \frac{LC_{pl}}{G_y}(T_x - T_{xa})$$

We see from the $H-T$ diagram that $T_{xa} = 80°F$ and $H_{yb} = 34$ Btu/lb D air

$$\text{humid vol. of inlet air} = 359\left(\frac{1}{28.9} + \frac{0.0205}{18}\right)\frac{540}{492} = \frac{14.1 \text{ ft}^3}{\text{lb D air}}$$

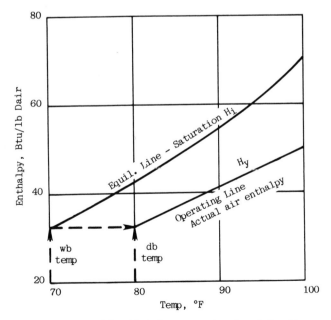

FIGURE M32a. Enthalpy-temperature for solution H.

So

$$G_y = \frac{750,000 \text{ ft}^3/\text{min}}{14.1 \text{ ft}^3/\text{lb D air}} = 53,191 \frac{\text{lb D air}}{\text{min}}$$

and

$$L = \left(\frac{5500 \text{ gal}}{\text{min}}\right)\left(61.8 \frac{\text{lb}}{\text{ft}^3}\right)\bigg/ 7.48 \frac{\text{gal}}{\text{ft}^3} = 45,441 \frac{\text{lb H}_2\text{O}}{\text{min}}, \qquad C_{pl} = 1\frac{\text{Btu}}{\text{lb}°\text{F}}$$

So

$$\frac{LC_{pl}}{G_y} = \frac{45,441 \times 1}{53,191} = 0.854 \frac{\text{Btu}}{\text{lb D air}°\text{F}}$$

$$H_y = 34 + 0.854(T - 80)$$

and we have plotted this, along with the saturation enthalpy in Figure M32a. We read data from this graph

T	80	85	90	95	100
H_i	43	49	55	63	71
H_y	34	38.3	42.5	46.8	51.1
$1/(H_i - H_y)$	0.1111	0.0935	0.08	0.0617	0.0503

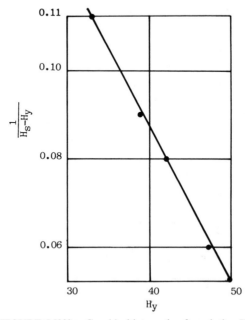

FIGURE M32b. Graphical integration for solution H.

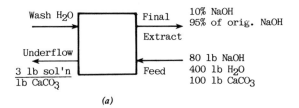

(a)

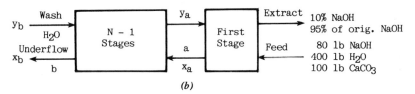

(b)

FIGURE M33. (a) Schematic for solution I; (b) modified schematic.

and plot $1/(H_i - H_y)$ versus H_y and graphically integrate Figure M32b to obtain $\int = 1.377$. But

$$\frac{C_s G'_y}{h_y a} = \left(\frac{0.255 \text{ Btu}}{\text{lb}°\text{F}}\right)\left(\frac{3.191 \times 10^6 \text{ lb D air}}{2025 \text{ ft}^2 \text{ hr}}\right)\left(\frac{\text{ft}^3 \text{ hr}°\text{F}}{10 \text{ Btu}}\right) = 40.2 \text{ ft}$$

so

$$Z = (40.2 \text{ ft}) \times (1.377) = 55.4 \text{ ft}$$

I. See Figure M33a. For a straight equilibrium line and a straight operating line, we use the Kremser equation. A constant underflow gives a straight operating line. But note that this equation cannot be used for the whole system if the solution entering with the unextracted solids is different from the underflow within the system, as in this case. Thus we must break the problem into two parts, as shown in Figure M33b.

Basis: Feed mixture.

$$\text{lb NaOH in extract} = 0.95 (80) = 76 \text{ lb}$$

but NaOH is 10% of the extract solution;

$$\text{lb extract} = 76 \text{ lb}/0.1 = 760 \text{ lb}$$

$$\text{lb final underflow} = (3 \text{ lb soln/lb CaCO}_3)100 \text{ lb CaCO}_3 + 100 \text{ lb CaCO}_3 = 400 \text{ lb}$$

Now the solution leaving a stage with the extract is the same concentration as the solution leaving a stage with the underflow. So point a has a solution containing 10% NaOH: $x_a = 0.1$ with 300-lb soln and 100-lb CaCO$_3$. To determine the wash water, we do a fluid balance around the total unit:

$$\begin{array}{cc} \text{in} & \text{out} \\ \text{wash water} + 400 + 80 = 760 + 300 & \text{or} \quad \text{wash water} = 580 \text{ lb} \end{array}$$

Kremser equation

$$N - 1 = \ln\left(\frac{y_b - x_b}{y_a - x_a}\right)\Big/\ln\left(\frac{y_b - y_a}{x_b - x_a}\right)$$

Final underflow will contain the unrecovered NaOH: 80 lb − 76 lb = 4 lb. It is contained in 300 lb of soln, so

$$x_b = 4/300 = 0.0133, \qquad y_b = 0$$

since wash water contains no NaOH.

To determine y_a, we do a NaOH balance around the first stage

$$\begin{array}{cc} \text{in} & \text{out} \\ 80 + 580y_a = 76 + 0.1(300) & \text{or} \qquad y_a = 0.0448 \end{array}$$

and

$$y_b = 0, \qquad x_a = 0.1, \qquad x_b = 0.0133$$

So

$$N - 1 = \ln\left(\frac{0 - 0.0133}{0.0448 - 0.1}\right)\Big/\ln\left(\frac{0 - 0.0448}{0.0133 - 0.1}\right) = \ln\left(\frac{-0.133}{-0.0552}\right)\Big/$$

$$\ln\left(\frac{-0.0448}{-0.0867}\right) = 2.16$$

Thus $N = 3.16$.

J. See Figure M34a. Slope of feed line $= q/(q - 1)$, where $q =$ moles of liquid flow in stripping section resulting from 1 mol of feed or $\frac{1}{2}$ feed vaporized or $\frac{1}{2}$ feed liquid, so

$$q = \tfrac{1}{2} \qquad \text{and} \qquad \text{slope} = \tfrac{1}{2}/(\tfrac{1}{2} - 1) = -1$$

Minimum reflux ratio is determined from slope of rectifying operating line through (y_d, x_d) and the intersection of feed with equilibrium curve. Slope of this line is $R_D/(R_D + 1)$, where $R_D = L/D$. Operating line goes through the point $x = 0.37$, $y = 0.63$ or

$$0.63 = \frac{R_D(0.37) + 0.97}{R_D + 1}$$

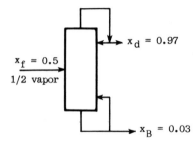

FIGURE M34a.   Schematic for solution J.

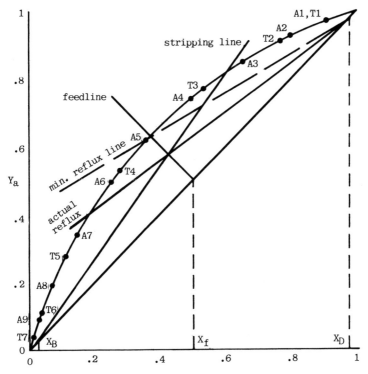

FIGURE M34*b*. McCabe–Thiele diagram for solution J.

(a)

$$R_D = 1.308 \text{ minimum reflux}$$

actual $R_D = 2 \times 1.308 = 2.616$,   actual slope $= 2.616/3.616 = 0.723$

(b) Theoretical stages using total reflux: At total reflux, the $y = x$ diagonal line is the operating line. From the construction shown in Figure M34*b*, we need $6+$, or 7, theoretical stages. This construction is indicated by the intersections with the equilibrium curve (T1, T2, etc.).

(c, d) We see from the construction using the actual reflux ratio of 2.616 that we need 9 theoretical stages, feeding onto the 5th theoretical plate from the top (A1, A2, etc.)

K. The minimum number of stages may be determined analytically with the Fenske equation

$$N_{min} = \frac{\ln \left[x_D (1 - x_B)/x_B (1 - x_D) \right]}{\ln \alpha} - 1$$

So with $x_D = 0.9$, $x_B = 0.1$, $\alpha = 1.75$, $N_{min} = 6.85$.

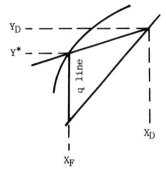

FIGURE M35. Equilibrium curve for solution K.

We have a saturated liquid feed, so $q = \infty$. Operating line for minimum reflux ratio goes through the points (y^*, x_F) and $(y_D = x_D, x_D)$ as indicated in Figure M35. So

$$\text{slope} = \frac{R_D}{R_D + 1} = \frac{y_D - y^*}{x_D - x_F} = \frac{x_D - y^*}{x_D - x_F}$$

or

$$(x_D - x_F)R_D = (x_D - y^*)(R_D + 1), \quad \text{or} \quad (x_D - x_F - x_D + y^*)R_D = x_D - y^*, \quad \text{or}$$

$$R_D = (x_D - y^*)/(y^* - x_F)$$

We need y^*, the equilibrium value at $x = x_F$. Now

$$\alpha = \frac{y(1 - x)}{x(1 - y)}$$

where for this case $y = y^*$, $x = x_F$ or

$$y^* = \frac{x_F}{1 + (\alpha - 1)x_F}$$

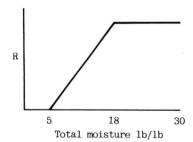

R

5   18   30
Total moisture lb/lb   **FIGURE M36a.** Drying data for solution L.

So

$$R_D = \left(x_D - \frac{\alpha x_F}{1 + (\alpha - 1)}\, x_F\right) \Big/ \left(\frac{\alpha x_F}{1 + (\alpha - 1)x_F} - x_F\right)$$

$$= \frac{x_D + (\alpha - 1)x_F x_D - x_F}{\alpha x_F - x_F - (\alpha - 1)x_F}$$

$$= \frac{x_D - x_F[x_D + \alpha(1 - x_D)]}{(\alpha - 1)x_F(1 - x_F)} = \frac{9 - 0.5[0.9 + 1.75(0.1)]}{0.75(0.5)(0.5)} = 1.93$$

L. Figure 36a uses total moisture; drying equations use free-moisture values, so

Total moisture $\left(\dfrac{\text{lb } H_2O}{\text{lb D solid}}\right)$	Free Moisture $\left(\dfrac{\text{lb } H_2O}{\text{lb D solid}}\right)$
30	$= 30 - 5 = 25$
18	$= 18 - 5 = 13$
12	$= 12 - 5 = 7$
8	$= 8 - 5 = 3$

We must use the given data to determine the rate-or-drying curve of Figure M36b.

Now

$$R = \frac{dX}{dt} \quad \text{or} \quad dt = \frac{dX}{R}$$

Constant rate:

$$t_c = \int_{13}^{25} \frac{dX}{R_c} = \frac{25 - 13}{R_c}$$

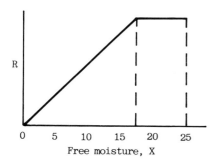

R

0   5   10   15   20   25

Free moisture, X

FIGURE M36b. Rate-of-drying curve: solution L.

Falling rate:

$$R = aX, \qquad \text{at } X = X_c, \quad R = R_C$$

so

$$a = \frac{R_c}{X_c} \qquad \text{or} \qquad R = \frac{R_C}{X_c} X$$

$$t_f = \left[\int_7^{13} dX \bigg/ \frac{R_C}{X_c} X = \frac{X_c}{R_c} \ln X \right]_7^{13}, \qquad t_f = \frac{13}{R_c} \ln \left(\frac{13}{7} \right)$$

So

$$5 = t_c + t_f = \frac{12}{R_c} + \frac{13}{R_c} \ln \left(\frac{13}{7} \right) = \frac{12 + 8.05}{R_c} = \frac{20.05}{R_c}$$

or

$$R_c = 4.01$$

Our equations for this system are

$$t_c = \frac{X - X_c}{R_c} = \frac{X - 13}{4.01}; \qquad t_f = \frac{X_c}{R_c} \ln \left(\frac{X_c}{X_{end}} \right) = \frac{13}{4.01} \ln \left(\frac{13}{X_{end}} \right)$$

So for $X = 25$ and $X_{end} = 3$,

$$t_c = \frac{25 - 13}{4.01} = 2.99 \text{ hr}; \qquad t_f = \frac{13}{4.01} \ln \left(\frac{13}{3} \right) = 4.75$$

and

$$\text{total time} = 2.99 \text{ hr} + 4.75 \text{ hr} = 7.74 \text{ hr}$$

HEAT TRANSFER

W. L. McCabe and J. C. Smith, *Unit Operations in Chemical Engineering*, 3rd ed., McGraw-Hill, New York, 1976.

F. Kreith, *Principles of Heat Transfer*, 3rd ed., Harper & Row, New York, 1973.

I. CONDUCTION IN SOLIDS

A. Introduction

1. Fourier's Law

The basic relation for heat flow by conduction is the proportionality between the rate of heat flow across a surface and the temperature gradient at the surface, or

$$\frac{dq}{dA} = -k\frac{\partial T}{\partial n}$$

where q = rate of heat flow normal to surface,
  A = area of surface,
  T = temperature,
  n = normal distance to surface, and
  k = proportionality constant.

2. Thermal Conductivity

The proportionality constant k is called the thermal conductivity. It is a physical property of the substance. It is independent of the temperature gradient, but it may be a function of the temperature itself.

B. Steady-State Conduction

1. Introduction

Consider a flat slab with a constant thermal conductivity. We look at the steady-state heat flow perpendicular to the slab. For this case, q will be constant. The direction normal to the slab will be called x. Fourier's law becomes

$$\frac{q}{A} = -k\frac{dT}{dx}$$

or

$$\int_{T_1}^{T_2} dT = -\frac{q}{kA}\int_{x_1}^{x_2} dx$$

or

$$\frac{q}{A} = k\frac{T_1 - T_2}{x_2 - x_1} = k\frac{\Delta T}{B}$$

where B = thickness of slab, $x_2 - x_1$, and
  ΔT = temperature drop across slab, $T_1 - T_2$

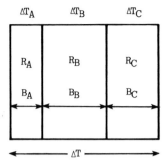

FIGURE H1. Compound resistances in series.

or, in resistance form

$$q = \Delta T/R$$

with

$$R = B/kA = \text{thermal resistance}$$

2. Compound Resistances in Series

Consider a flat wall made of a series of layers as shown in Figure H1. For the first material (or layer), R_A is the resistance, B_A is the thickness, and T_A is the temperature drop across B_A. We can write

$$\Delta T_A = q_A \frac{B_A}{k_A A}, \qquad \Delta T_B = q_B \frac{B_B}{k_B A}, \qquad \Delta T_C = q_C \frac{B_C}{k_C A}$$

$$\Delta T_A + \Delta T_B + \Delta T_C = \Delta T = \frac{q_A B_A}{A k_A} + \frac{q_B B_B}{A k_B} + \frac{q_C B_C}{A k_C}$$

and, since $q_A = q_B = q_C = q$,

$$q = \Delta T \bigg/ \left(\frac{B_A}{k_A A} + \frac{B_B}{k_B A} + \frac{B_C}{k_C A} \right) = \frac{\Delta T}{R_A + R_B + R_C} = \frac{\Delta T}{R}$$

3. Heat Flow through a Cylinder

Consider the concentric tube shown in Figure H2. Heat is flowing from the inside to the outside. Note that the area perpendicular to this flow changes with r, that is,

$$q = -k \frac{dT}{dr} 2\pi r L$$

$$L = \text{length of tube}$$

Integrating

$$\int_{r_i}^{r_o} \frac{dr}{r} = -\frac{2\pi k L}{q} \int_{T_i}^{T_o} dT$$

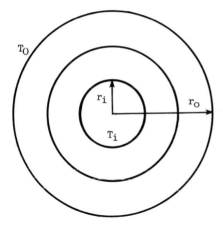

FIGURE H2.   Concentric tubes.

or

$$q = \frac{k2\pi L(T_i - T_o)}{\ln (r_o/r_i)} = \frac{kA_L(T_i - T_o)}{r_o - r_i}$$

where the last form is the way we wish to write the equation. So, in order to use this form, we must have

$$A_L = \frac{2\pi L(r_o - r_i)}{\ln (r_o/r_i)}$$

We have just defined the logarithmic mean radius.

C.   Unsteady-State Heat Conduction

Consider the earlier flat slab. We now look at it, not in the steady state, but as time varies. The conduction equation becomes

$$\frac{\partial T}{\partial t} = \frac{k}{\rho C_p} \frac{\partial^2 T}{\partial x^2} = \alpha \frac{\partial^2 T}{\partial x^2}$$

Because the solution to this equation is a series of terms, charts (such as those found in Kreith) are convenient.

II.   HEAT FLOW IN FLUIDS

A.   Introduction

Consider Figure H3 in which a liquid is being heated by a condensing vapor. The inlet and outlet fluid temperatures are T_{ca} and T_{cb}. The constant vapor temperature is T_h. At some length from the entrance, the liquid temperature is T_c

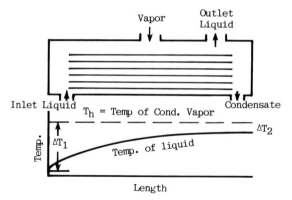

FIGURE H3.   Heat exchanger with condensing vapor.

and so the local difference is $T_h - T_c$. This is called a point temperature difference, ΔT. At the inlet, the point temperature difference is $T_h - T_{ca} = \Delta T_1$; at the
exit $T_h - T_{cb} = \Delta T_2$. The terminal point temperature differences ΔT_1 and ΔT_2
are called approaches. The change in the liquid temperature $T_{cb} - T_{ca}$ is called
the temperature range.

Single pass systems can be either parallel or countercurrent. Countercurrent
almost always is the preferred method.

B.   Energy Balances

1.   *In Heat Exchangers*

The usual energy balance is

$$q = \dot{m}(H_b - H_a)$$

where $q = Q/t =$ rate of heat transfer into the stream,
  $\dot{m} =$ flow rate of the stream, and
  $H_a, H_b =$ enthalpy per unit mass, in and out.

If we assume two fluid streams exchanging heat, with no loss to the surroundings,

$$q_h = \dot{m}_h(H_{hb} - H_{ha}), \qquad q_c = \dot{m}_c(H_{cb} - H_{ca})$$

h = hot stream, c = cold stream.

Note the sign of q_c is positive, while that of q_h is negative, that is, $q_c = -q_h$.
So, the overall enthalpy balance is

$$\dot{m}_h(H_{ha} - H_{hb}) = \dot{m}_c(H_{cb} - H_{ca}) = q$$

If we assume constant specific heats,

$$\dot{m}_h C_{ph}(T_{ha} - T_{hb}) = \dot{m}_c C_{pc}(T_{cb} - t_{ca}) = q$$

2. In Total Condensers

With a total condenser, the hot stream gives up its heat of vaporization only, that is,

$$\dot{m}_h \lambda = \dot{m}_c C_{pc}(T_{cb} - T_{ca}) = q$$

If the hot stream also is subcooled

$$\dot{m}_h[\lambda + C_{ph}(T_h - T_{hb})] = \dot{m}_c C_{pc}(T_{cb} - T_{ca}) = q$$

C.   Rate of Heat Transfer

1.   Heat Flux

Our heat transfer calculations will be based upon the area of the heating surface. It is usually expressed as British thermal units (Btu) per hour per square foot or watts per square meter. This rate of heat transfer per unit area is called a heat flux. For tube systems, it can be based either upon the outside tube area or upon the inside tube area.

2.   Average Temperature of Stream

Bear in mind that there is usually a temperature difference across the cross section of the fluid. But we use the mixing cup or average temperature of the cross section.

3.   Overall Heat Transfer Coefficient

Analogous to the Fourier law of continuous mediums, we can write for the flux dq/dA,

$$\frac{dq}{dA} = U\Delta T = U(T_h - T_c)$$

where the proportionality factor U is called the local overall heat transfer coefficient. But note that if A is taken to be the outside tube area A_o, then U becomes U_o, the coefficient based upon that area; likewise for A_i and U_i. Now both ΔT and dq are independent of the area chosen, so

$$\frac{U_o}{U_i} = \frac{dA_i}{dA_o} = \frac{D_i}{D_o}, \qquad D = \text{tube diameters}$$

4. Log Mean Temperature Difference

If we assume that the overall heat transfer coefficient is constant, the flux equation can be integrated over the total heat transfer surface to get

$$q_T = U A_T \overline{\Delta T_L}$$

$$\overline{\Delta T_L} = \frac{\Delta T_2 - \Delta T_1}{\ln (\Delta T_2 / \Delta T_1)} = \text{LMTD}$$

5. Variable Overall Heat Transfer Coefficient

If we assume that U varies linearly with temperature drop, a more accurate equation is

$$q_T = A_T \frac{U_2 \Delta T_1 - U_1 \Delta T_2}{\ln (U_2 \Delta T_1 / U_1 \Delta T_2)}$$

6. Overall Coefficients from Individual Coefficients

The overall heat transferred depends upon heat transfer in the hot stream, in the tube wall, and in the cold stream. Thus the overall coefficient is a function of individual coefficients:

$$U_o = 1 \left/ \left(\frac{D_o}{D_i h_i} + \frac{x_w}{k} \frac{D_o}{D_L} + \frac{1}{h_o} \right) \right.$$

if A_o is the area, and

$$U_i = 1 \left/ \left(\frac{1}{h_i} + \frac{x_w}{k} \frac{D_i}{D_L} + \frac{D_i}{D_o h_o} \right) \right.$$

if A_i is the area used, where

h_i, h_o = inside and outside heat transfer coefficient,
D_i, D_o = inside and outside tube diameters,
x_w = tube wall thickness,
D_L = log mean diameter of tube, and
k = thermal conductivity of wall.

7. Resistance Form of Overall Coefficient

The overall coefficient can be written in resistance form

$$\frac{1}{U_o} = \frac{D_o}{D_i h_i} + \frac{x_w}{k} \frac{D_o}{D_L} + \frac{1}{h_o}$$

8. Fouling Factors

While in service, heat transfer surfaces tend to get dirty or fouled. This is taken into account by using fouling factors h_{di} or h_{do}, depending upon whether dirt is on the inside or outside of the tube. They may be incorporated into the overall coefficients in the following way:

$$\frac{1}{U_o} = \frac{D_o}{D_i h_{di}} + \frac{D_o}{D_i h_i} + \frac{x_w D_o}{kD_L} + \frac{1}{h_o} + \frac{1}{h_{do}}$$

$$\frac{1}{U_i} = \frac{1}{h_{di}} + \frac{1}{h_i} + \frac{k_w D_i}{kD_L} + \frac{D_i}{D_o h_o} + \frac{D_i}{D_o h_{do}}$$

III. HEAT TRANSFER TO FLUIDS—NO PHASE CHANGES

A. Forced Convection—Laminar Flow

1. Introduction

With laminar flow, heat transfer occurs only by conduction. This type of problem may be solved using mathematical analysis. An advanced text on fluid flow and heat transfer will give the details; only the results will be presented here.

2. Heat Transfer to a Flat Plate

The equation for heat transfer to a flat plate (shown in Figure H4) is

$$N_{Nu,x} = \frac{0.332 N_{Pr}^{1/3} N_{Re,x}^{1/2}}{\{1 - (x_o/x)^{3/4}\}^{1/3}}$$

where $N_{Nu,x} = h_x x/k$ = Nusselt number,

$$N_{Pr} = \frac{C_p \mu}{k}$$ = Prandtl number, and

$$N_{Re,x} = \frac{U_o x \rho}{\mu}$$ = Reynolds number.

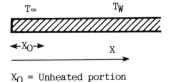

T_∞ T_W

←X_O→   X

X_O = Unheated portion of plate

FIGURE H4. Heat transfer to a flat plate.

This equation gives h_x, the local value of h. Analytically it can be shown that the average h, over the whole length x, is given by

$$h = 2h_L$$

where h_L is h_x evaluated at $x = L$.

3. Graetz and Peclet Numbers

Dimensionless groups are commonly used in heat transfer to fluids. The Graetz number is defined as

$$N_{Gz} = \frac{\dot{m}C_p}{kL} = \frac{\pi}{4}\rho \frac{\bar{V}C_p D^2}{kL}$$

The Peclet number is defined to be the product of the Reynolds number and the Prandtl number

$$N_{Pe} = N_{Re}N_{Pr} = \frac{D\bar{V}\rho}{\mu}\frac{C_p\mu}{k} = \rho\frac{\bar{V}C_p D}{k}$$

B.   Forced Convection—Turbulent Flow

1.   Introduction

The most important heat transfer situation is heat flow to a turbulent fluid flowing in a tube. Turbulent flow has Reynolds numbers greater than 2100. The first results obtained were empirical correlations using dimensional analysis; later results have had a more theoretical basis.

2.   Dimensional Analysis Results

Most turbulent-flow forced-convection heat transfer data can be correlated by equations of the form

$$\frac{h}{C_p G} = \phi\left(\frac{DG}{\mu}, \frac{C_p\mu}{k}\right)$$

or

$$N_{St} = \phi(N_{Re}, N_{Pr}), \qquad N_{St} = \text{Stanton number}$$

For sharp edged entrances, the equation is

$$\frac{h}{C_p G}\left(\frac{C_p\mu}{k}\right)^{2/3}\left(\frac{\mu_w}{\mu}\right)^{0.14} = \frac{0.023[1 + (D/L)^{0.7}]}{(DG/\mu)^{0.2}}$$

All physical property data are evaluated at the bulk temperature T, except μ_w which is evaluated at the wall temperature. To estimate the wall temperature, an iterative calculation is performed using the resistance equations.

3. Theoretical Results

Some theoretical results may be obtained using the Reynolds analogy—the ratio of the momentum diffusivity to the thermal diffusivity is equal to the ratio of the momentum eddy diffusivity to the thermal eddy diffusivity. It is theoretically expressed as

$$(h/C_p G)N_{Pr} = f/2$$

The following experimentally determined equation is more accurate:

$$\frac{h}{C_p G} N_{Pr}^{2/3}\left(\frac{\mu_w}{\mu}\right)^{0.14} = \frac{f}{2}$$

$$f = 0.046(DG/\mu)^{-0.2}$$

More accurate still is the form

$$N_{St} = \frac{f/2}{\phi(N_{Pr})}$$

but you must know $\phi(N_{Pr})$ for various cases.

C. Forced Convection outside Tubes

1. Introduction

Heat flow in forced convection outside tubes differs from that inside tubes due to differences in mechanisms. No form drag exists inside the tubes and all friction is wall friction.

2. Flow Normal to a Single Tube

The same variables affecting heat transfer within tubes affect heat transfer outside, but the constants are different. The form is

$$\frac{h_o D_o}{k} = \psi_o\left(\frac{D_o G}{\mu}, \frac{C_p \mu}{k}\right)$$

For heating and cooling fluids flowing normal to single cylinders, use

$$\frac{h_o D_o}{k_f}\left(\frac{C_p \mu_f}{k_f}\right)^{-0.3} = 0.35 + 0.56\left(\frac{D_o G}{\mu_f}\right)^{0.52}$$

where f means evaluated at the film temperature,

T_f = film temperature = $(T_w + \bar{T})/2$, with T_w = wall temperature, and
$\bar{T}$ = bulk temperature.

3.   Flow past a Single Sphere

The equation for this case is

$$\frac{h_o D_o}{k_f} = 2.0 + 0.60 \left(\frac{D_p G}{\mu_f}\right)^{0.50} \left(\frac{C_p \mu_f}{k_f}\right)^{1/3}$$

with D_p = sphere diameter.

D.   Natural Convection

A fluid, at rest, in contact with a hot vertical plate will be warmer near the plate. Since the density of a hot fluid is less than that of a cold fluid, this buoyancy effect will cause the hot fluid to rise and be replaced with a cooler fluid. A velocity gradient will be formed near the plate. This is natural convection.

For single horizontal cylinders, the equation is of the form

$$\frac{h D_o}{k_f} = \phi\left(\frac{C_p \mu_f}{k_f}, \frac{D_o^3 \rho_f^2 \beta g \Delta T_o}{\mu_f^2}\right)$$

where the last term is the Grashof number N_{Gr} and

β = coefficient of thermal expansion of the fluid
ΔT_o = average difference in temperature between the bulk fluid
        and the cylinder wall.

Mathematically, the coefficient of thermal expansion is

$$\beta = \frac{1}{v}\left(\frac{\partial v}{\partial T}\right)_P, \qquad v = \text{specific volume}$$

For liquids, you can use

$$\beta = \frac{\rho_1 - \rho_2}{\rho_a(T_2 - T_1)}, \qquad \rho_a = \frac{\rho_1 + \rho_2}{2}$$

For ideal gases

$$\beta = 1/T$$

For $\log_{10} N_{Gr} N_{Pr} > 4$

$$N_{Nu} = 0.53(N_{Gr} N_{Pr})_f^{0.25}$$

IV.   HEAT TRANSFER TO FLUIDS—PHASE CHANGES

A.   Condensing Vapors

1.   Introduction

The condensation of vapors upon the surfaces of tubes which are cooler than the vapor condensing temperature is important commercially. The important areas will be discussed.

There are two ways in which a vapor may condense upon a cold surface: dropwise and filmwise. The average coefficient for dropwise condensation may be five to eight times that for filmwise condensation. Some generalizations are

- Filmwise condensation occurs if both the fluid and the tube are clean.
- Dropwise condensation occurs only when the surface is contaminated.
- A monomolecular film is sufficient to cause dropwise condensation.
- Substances that merely prevent surface wetting are ineffective for causing dropwise condensation.

To be conservative, use filmwise condensation for design.

2. Coefficients for Filmwise Condensation

These equations have a strong theoretical basis. For vertical tubes, the local heat transfer coefficient at a distance L from the surface top is

$$h_x = k_f \left(\frac{\rho_f^2 g}{3 \Gamma \mu_f} \right)^{1/3}$$

or equivalently by

$$h = \frac{\lambda \Gamma_b}{L_T \Delta T_o}$$

for an average coefficient, where

$\Gamma = \dot{m}$ divided by wetted perimeter,
$\lambda = $ heat of vaporization,
$L_T = $ total tube length, and
$\Gamma_b = $ condensate loading at bottom of tube.

To remove the temperature difference, some manipulations are required, obtaining

$$h \left(\frac{\mu_f^2}{k_f^3 \rho_f^2 g} \right)^{1/3} = 1.47 \left(\frac{4 \Gamma_b}{\mu_f} \right)^{-1/3}$$

The physical property data are evaluated at T_f where

$$T_f = T_h - \frac{3(T_h - T_w)}{4}$$

with $T_h = $ temperature of condensing vapor and
$T_w = $ wall temperature.

This equation is equivalent to

$$h = 0.943 \left(\frac{k_f^3 \rho_f^2 g \lambda}{\Delta T_o L \mu_f} \right)^{1/4}$$

where $\Delta T_o = T_h - T_w$,

For horizontal tubes, use either

$$h\left(\frac{\mu_f^2}{k_f^3 \rho_f^2 g}\right)^{1/3} = 1.51\left(\frac{4\Gamma'}{\mu_f}\right)^{-1/3}$$

or

$$h = 0.725\left(\frac{k_f^3 \rho_f^2 g \lambda}{\Delta T_o D_o \mu_f}\right)^{1/4}$$

where Γ' is the condensate loading per unit length of tube.

3. Practical Use of Equations

These theoretical equations work well for single tubes. Theoretically for a bank of horizontal tubes, the average loading per tube could be used

$$\Gamma_s' = \frac{\dot{m}_T}{(LN)}$$

where $\dot{m}_T$ = total condensate flow rate for stack,
L = length of one tube, and
N = number of tubes in the stack.

Because of splashing, better results are obtained by using

$$\Gamma_s' = \dot{m}_T/(LN^{2/3})$$

4. Superheated Vapors

If the vapor is superheated, both the sensible heat (of the superheat) and the heat of condensation must be transferred through the cooling surface. When the vapor is highly superheated and the outlet temperature of the cooling fluid is close to that of condensation, the tube wall temperature may be greater than the saturation temperature and condensation will not occur. In this case, it may be necessary to consider the equipment as two separate parts: a desuperheater (just a gas cooler) and then a condenser.

B. Boiling Liquids

1. Introduction

When boiling is caused by a hot immersed surface, the temperature of the liquid is the same as the boiling point of the liquid under the pressure in the equipment. Vapor bubbles are generated at the heating surface, rise through the liquid, and leave the liquid surface. This is boiling of saturated liquid.

In some cases, the liquid temperature is below the boiling point but the heating surface temperature is above the boiling point. Vapor bubbles formed on the heating surface are absorbed by the liquid. This is subcooled boiling.

2. *Boiling of Saturated Liquid*

The general equation is given by

$$q/A = a\Delta T^b$$

but a and b change depending upon the liquid and the temperature range. Experimental data are required. McCabe and Smith (pp. 363–366) report the available equations, such as they are.

3. *Subcooled Boiling*

Very little is known about this topic.

V. RADIATION HEAT TRANSFER

A. Introduction

Radiation is energy streaming through empty space at the speed of light. All materials at temperature above absolute zero emit radiation. Thermal radiation is radiation that is the result of temperature only. Since radiation moves through space in straight lines, only substances in sight of a radiating body can intercept radiation from the body. The fraction of radiation hitting a body that is reflected is called the reflectivity ρ. The fraction absorbed is called the absorptivity α, while the fraction transmitted is called the transmissivity τ. The sum of these three fractions must be 1, or

$$\alpha + \rho + \tau = 1$$

Of course, the maximum possible absorptivity is 1, and this is obtained only if the body absorbs all incident radiation and transmits or reflects none. Such a body is called a blackbody.

B. Emission of Radiation

1. Introduction

The radiation emitted by a material is independent of radiation being emitted by other bodies in sight of, or in contact with it. Of course the net energy is the difference between the energy emitted and that absorbed by it. With different temperature bodies in sight of each other within an enclosure, the hotter bodies lose energy by emission faster than they receive energy by absorption from cooler bodies. Eventually the process reaches equilibrium when all bodies reach the same temperature.

2. Radiation Wavelength

Radiation of a single wavelength is called monochromatic, but actual radiation beams consist of many monochromatic beams. The higher the temperature of the body radiating, the shorter is the predominant wavelength emitted by it.

The monochromatic energy emitted by a surface depends upon the surface temperature and wavelength of the radiation. The monochromatic radiating power W is the monochromatic radiation emitted from unit area in unit time. For the entire spectrum, the total radiating power W is $\int_0^\infty W_\lambda \, d\lambda$.

3. Blackbody Radiation and Emissivity

A blackbody has the maximum attainable emissive power at any given temperature; it is the standard. Thus the ratio of the total emissive power W of a body to that of a blackbody, W_b, is defined to be the emissivity ε of the bodies. So

$$\varepsilon = W/W_b$$

Likewise, the monochromatic emissivity ε_λ is defined by

$$\varepsilon_\lambda = W_\lambda/W_{b,\lambda}$$

A gray body is defined to be a body whose ε_λ is the same for all wavelengths λ.

4. Practical Source of Blackbody Radiation

There are no actual blackbodies, but carbon black approaches it. There is a basic blackbody relationship, the Stefan–Boltzmann law,

$$W_b = \sigma T^4, \qquad \sigma = 0.1713 \times 10^{-8} \text{ Btu/ft}^2 \text{ hr } ^\circ\text{R}^4$$

with σ a universal constant. Planck's law gives the energy distribution in the spectrum of a blackbody

$$W_{b,\lambda} = \frac{C_1 \lambda^{-5}}{e^{C_2/\lambda T} - 1}, \qquad \begin{aligned} C_1 &= 3.742 \times 10^{-16} \text{ Wm}^2 \\ C_2 &= 1.439 \text{ cmK} \end{aligned}$$

At any temperature T, the maximum monochromatic radiating power is obtained at a wavelength λ_{max} given by Wien's displacement law:

$$T\lambda_{max} = 5200; \qquad T = {}^\circ\text{R}, \qquad \lambda_{max} = \text{micrometers}$$

C. Absorption of Radiation by Opaque Solids

1. Introduction

A definite fraction ρ of radiation falling on a solid body may be reflected and the remaining fraction $1 - \rho$ enters the solid and is either transmitted or absorbed. In most solids the transmissivity τ is negligible. In fact, most of the absorbed radiation in an opaque solid is absorbed in a thin layer near the surface. Then the heat generated by the absorption can flow into the remaining solid by conduction.

2. Reflectivity and Absorptivity of Opaque Solids

An opaque solid's reflectivity depends mainly upon temperature and the type of surface. Reflection is usually either specular or diffuse. Specular reflection makes a definite angle with the surface, usually occurring on polished metals. Diffuse reflection is reflected in all directions, usually the result of a rough surface.

3. Kirchhoff's Law

At temperature equilibrium, the ratio of the total radiating power of a body to the absorptivity of that body depends only upon the body temperature. So for two bodies 1 and 2,

$$\frac{W_1}{\alpha_1} = \frac{W_2}{\alpha_2}$$

with W = total radiating power and
α = absorptivity.

If one of the bodies is a blackbody

$$\varepsilon = W/W_b = \alpha$$

that is, emissivity and absorptivity are equal when a body is at temperature equilibrium with the surroundings.

D. Radiation between Surfaces

1. Introduction

The total radiation from a unit area of an opaque body of area A_1, emissivity ε_1, and absolute temperature T is given by

$$q/A_1 = \sigma\varepsilon_1 T_1^4$$

But, in the usual situation, the energy emitted by a body is intercepted by other bodies (or substances) in sight of the body. And these substances are also radiators and their radiation falls on the original body. So the net loss by radiation is less than our equation indicates.

Consider the simplest case of two parallel surfaces 1 and 2 with both surfaces black. The energy emitted by surface 1 is σT_1^4. But all radiation from each of the surfaces falls on the other surface and is completely absorbed. Thus the net loss of energy by the first surface and the net gain of energy by the second is given by

$$\sigma T_1^4 - \sigma T_2^4 = \sigma(T_1^4 - T_2^4) \qquad \text{if } T_1 > T_2$$

Usual real world situations are more complex, since one or both surfaces usually see other surfaces and actual surfaces are rarely black.

2. Angle of Vision

Interception of radiation from an area element of a surface by another surface of finite size can be seen in terms of the angle of vision, the solid angle subtended by the finite surface at the radiating element.

3. Square of the Distance Effect

The energy from a small surface dA_1 that is intercepted by a large one depends only upon the angle of vision and is independent of the distance between the surfaces. But the energy received per unit area of the receiving surface is inversely proportional to the square of the distance between the surfaces, or

$$dI = (W_1/\pi r^2)\, dA_1 \cos \phi$$

where I = rate of energy received per unit area of the receiving surface, and the angle ϕ is determined as shown in Figure H5.

4. Quantitative Calculation of Radiation between Black Surfaces

The net rate of energy transfer dq_{12} between two area elements dA_1 and dA_2 is given by

$$dq_{12} = \sigma \frac{\cos \phi_1 \cos \phi_2 \, dA_1 \, dA_2}{\pi r^2} (T_1^4 - T_2^4)$$

where ϕ_1 = angle between dA_1 normal and line of sight,
ϕ_2 = angle between dA_2 normal and line of sight, and
r = line of sight distance between areas.

The problem arises when we attempt to do the double integration for different geometries. In any case, the resulting equation would be

$$q_{12} = \sigma A F(T_1^4 - T_2^4)$$

where q_{12} = net radiation between surfaces 1 and 2,
A = area of either of the surfaces, arbitrary choice, and
F = dimensionless geometric factor.

FIGURE H5. Angle of vision.

The F factor is called the angle factor or the view factor. If surface A_1 is chosen,

$$q_{12} = \sigma A_1 F_{12}(T_1^4 - T_2^4)$$

If surface A_2 is chosen,

$$q_{12} = \sigma A_2 F_{21}(T_1^4 - T_2^4)$$

Thus

$$A_1 F_{12} = A_2 F_{21}$$

We can consider F_{12} as the fraction of radiation leaving area A_1 that is intercepted by area A_2. If surface A_1 sees only A_2, then $F_{12} = 1$. If surface A_1 sees several surfaces,

$$F_{11} + F_{12} + F_{13} + \cdots = 1$$

where F_{11} occurs in case area 1 can see area 1. Regardless, the net radiation associated with an F_{11} factor is zero.

Many view factors have been calculated. See McCabe and Smith, pp. 384–386.

5. Nonblack Surfaces

If we have gray surfaces, the preceding equation becomes

$$q_{12} = \sigma A_1 \mathscr{T}_{12}(T_1^4 - T_2^4) = \sigma A_2 \mathscr{T}_{21}(T_1^4 - T_2^4)$$

where $\mathscr{T}_{12}$ and $\mathscr{T}_{21}$ are overall interchange factors and depend upon ε_1 and ε_2. There are several other cases that may be developed analytically; see McCabe and Smith, pp. 387–389.

E. Radiation to Semitransparent Materials

1. Introduction

Many industrially important substances are transparent to the passage of radiant energy. Glasses and many gases are in this category.

2. Radiation to Absorbing Gases

The fraction of incident radiation absorbed by a given amount of gas depends upon the radiation path length and on the number of molecules met during the passage. Designing gas fired heaters, because of this, is a difficult task. For methods, consult a heat transfer text.

VI. HEAT EXCHANGERS

A. Heat Transfer Coefficients in Shell and Tube Exchangers

Tube side heat transfer coefficients cause no difficulty. But because of the shell side flow patterns, shell side, or outside, coefficients must be approached differently.

The mass velocity G_b parallel to the tubes is the mass flow rate divided by the free area for flow in the baffle window S_b (the portion of the shell cross section not occupied by the baffle). This area is calculated by

$$S_b = f_b \frac{\pi D_s^2}{4} - N_b \frac{\pi D_o^2}{4}$$

where f_b = fraction of shell occupied by baffle window (usually 0.1955),
  D_s = shell inside diameter,
  D_o = tube outside diameter, and
  N_b = number of tubes in baffle window.

For shell side crossflow, the mass velocity G_c is based upon the area S_c for transverse flow between the tubes in the row nearest the shell centerline. It can be estimated from

$$S_c = PD_s\left(1 - \frac{D_o}{p}\right)$$

where p = center-to-center distance between tubes and
  P = baffle pitch.

With the area determined, the mass velocity is calculated from

$$G_b = \frac{\dot{m}}{S_b} \qquad \text{for parallel shell flow}$$

and

$$G_c = \frac{\dot{m}}{S_c} \qquad \text{for crossflow in the shell}$$

The mass velocities are used with the Donohue equation

$$\frac{h_o D_o}{k} = 0.2\left(\frac{D_o G_e}{\mu}\right)^{0.6}\left(\frac{C_p \mu}{k}\right)^{0.33}\left(\frac{\mu}{\mu_w}\right)^{0.14}$$

where $G_e = \sqrt{G_b G_c}$

B. LMTD Correction for Crossflow

When a fluid flows perpendicular to the tube bank, the usual log mean temperature difference (LMTD) only applies if either the inside or the outside temperature is constant. otherwise corrections must be used. These are obtained from figures such as those in McCabe and Smith, pp. 405–407, where

$$Z = \frac{T_{ha} - T_{hb}}{T_{cb} - T_{ca}}, \qquad \eta_H = \frac{T_{cb} - T_{ca}}{T_{ha} - T_{ca}}$$

are used to give F_G to multiply the usual LMTD by, with

T_{ha}, T_{hb} = hot fluid, inlet and outlet temperatures
T_{ca}, T_{cb} = cold fluid, inlet and outlet temperatures

VII. EVAPORATORS

A. Introduction

To concentrate a solution of a nonvolatile solute in a volatile solvent, evaporation is usually preferred. For most evaporations the solvent is water. In an actual evaporator, the liquid head over the tubes affects performance. Entrainment of solids and the presence of noncondensables in the vapor stream can be a practical problem. In preliminary designs these effects are ignored. Heat losses are always assumed negligible as are the superheat in the stream and the subcooling in the condensate.

B. Evaporator Capacity

Capacity is defined as the amount of water vaporized per hour. To determine capacity, the rate of heat transfer through the heating surface of the evaporator must be known. The evaporator functions as a heat exchanger so the heat transfer rate is given by

$$q = UA\,\Delta T$$

The capacity is proportional to q if the evaporator feed is at the boiling temperature corresponding to the pressure in the vapor space. But, if the feed is colder, some of the q must be used to heat the feed to the boiling point. This reduces the heat available for evaporation and the capacity decreases. Increased capacity occurs if the feed enters above the boiling point.

1. Boiling-Point Elevation

At a given pressure, the boiling point of aqueous solutions is higher than that of pure water. This temperature difference is called the boiling-point elevation of the solution. It must be subtracted from the temperature drop that is predicted from the steam tables. Duhring's rule—the boiling point of a given solution is a linear function of the boiling point of pure water at the same pressure—is used for concentrated aqueous solutions.

C. Evaporator Economy

Economy is the amount of solution water vaporized per pound of steam fed to the evaporator. If the water vapor from one evaporator becomes the heating medium (steam) for a second evaporator, the economy increases. How much it improves can be determined by enthalpy balances.

D. Single-Effect Evaporation

Consider the evaporator shown in Figure H6. If we neglect steam superheat and condensate subcooling the enthalpy balance for the steam chestside is

$$q = \dot{m}_s(H_c - H_s) = -\dot{m}_s\lambda_s$$

where H_c = enthalpy of condensate,
$\quad H_s$ = enthalpy of steam,
$\quad \dot{m}_s$ = steam flow rate,
$\quad q_s$ = heat transfer rate through heating surface to steam, and
$\quad \lambda_s$ = latent heat of condensation of steam.

For the liquor side, the enthalpy balance is

$$q = (\dot{m}_f - \dot{m})H_v - \dot{m}_f H_f + \dot{m}H$$

where H = enthalpy of thick liquor,
$\quad H_f$ = enthalpy of feed liquor,
$\quad H_v$ = enthalpy of vapor, and
$\quad q$ = heat transfer rate from heating surface to liquid.

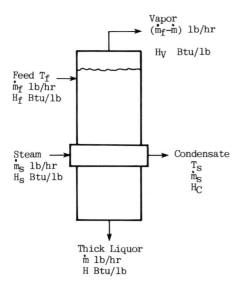

FIGURE H6. Evaporator.

If we assume no heat loss, $q = -q_s$ or

$$q = \dot{m}_s \lambda_s = (\dot{m}_f - \dot{m})H_v - \dot{m}_f H_f + \dot{m}H$$

Bear in mind that H and H_f depend upon both temperature and concentration. Perry's *Chemical Engineers' Handbook* has enthalpy concentration diagrams for many aqueous systems. Once the enthalpy balance (and material balances) has been completed, we note that

$$\text{capacity} = \dot{m}_f - \dot{m}$$
$$\text{economy} = (\dot{m}_f - \dot{m})/\dot{m}_s$$

E. Multiple-Effect Evaporation

In multiple-effect evaporation, the first effect is heated with steam, but the subsequent effects are heated by the prior effect's vapor stream. But each effect uses the single-effect material and enthalpy balances. Common practice dictates that all of the effects have the same heating area. Trial-and-error calculations are usually required. McCabe and Smith recommend the following calculation procedure:

- Assume unknown boiling temperatures.
- Using enthalpy balances, find steam and liquor flows between effects.
- Determine the needed heating surface A.
- If the areas are not equal, estimate new boiling temperatures.

To help in this procedure, it can be shown that in a multiple-effect evaporator train, the temperature drops are inversely proportional to the heat transfer coefficients.

PROBLEMS

A. We wish to design a kiln to withstand a temperature of 2000°F while limiting the heat loss to 250 Btu/hr ft² with an outside temperature of 100°F. We have available the following types of brick:

		$k\left(\dfrac{\text{Btu ft}}{\text{hr ft}^2\,{}^\circ\text{F}}\right)$	Thickness (in.)	Max Allowable Temp (°F)
(a)	Fireclay	0.90	4.5	—
(b)	Insulating	0.12	3.0	1800°
(c)	Building	0.40	4.0	300°

What will be the minimum wall thickness and what is the actual heat loss through the wall?

B. We wish to build an economical coldroom 20 ft square and 10 ft high. The outside temperature is 80°F and the room is to be maintained at 0°F. The boss has suggested parallel wooden outer and inner walls separated by a 6-in. air space. What will be the heat loss through each wall? Our bright new engineer suggests that the parallel outer and inner wall be divided in half by a thin (0.001-in.) sheet of aluminum foil. Should we fire her?

C. To cool a large slab of glass 4 in. thick from 400 to 100°F (center temperature), we put it into a constant temperature bath at 50°F. How long must the glass remain in the bath if:

$$\rho = 160 \text{ lb/ft}^3, \quad C_p = 0.18 \text{ Btu/lb°F}, \quad k = 0.36 \text{ Btu/hr ft°F}, \quad h \gg (k/r_m)?$$

D. We wish to heat 5000 lb/hr ft² (based on minimum free area) of 50°F air by passing it across an available tube bank that contains $\frac{3}{4}$ in. OD tubes on $1\frac{1}{2}$ in. centers with square in-line arrangements of 12 rows in each direction. If the tube walls are at 200°F, what will be the outlet temperature of the air?

E. Forty-five pounds per minute of water is heated from 60 to 180°F in a $\frac{5}{8}$ in., 16 BWG (Birmingham Wire Gage) heat exchanger tube. If the steam condensing, at 260°F, on the outside of the tube gives an h_o of 1000 Btu/hr ft² °F, calculate the inside coefficient, the overall coefficient, and the tube length required. After a half a year of operation, the overall heat transfer coefficient decreased by half. What is the inside scale coefficient?

F. A $1\frac{1}{4}$-in. Schedule 40 steel pipe is insulated with a 2-in. layer of asbestos covered by a 3-in. layer of 85% magnesia. If the temperature inside the pipe wall is 300°F and the atmospheric temperature is 80°F, what is the heat loss per foot of pipe and what is the temperature between the two insulations?

$$k_{\text{steel}} = 26, \qquad k_{\text{asb}} = 0.087, \qquad k_{85\%} = 0.034 \text{ Btu/hr ft °F}$$

G. Our old 1–2 heat exchanger heats 55 gpm of water from 60 to 130°F using 210 gpm of shell side hot water entering at 240°F. We estimate a dirt factor of 0.002 and a dirty U_o of 300. What were the two outlet temperatures when the exchanger was first installed?

H. We wish to concentrate a colloidal solution from 15 to 60% solids. The specific heat of the feed, entering at 50°F is 0.93. The pressure of the saturated steam to the evaporator is 10 psia; the condenser pressure is 5-in. Hg abs. The steam chest has an overall heat transfer coefficient of 200 Btu/hr ft² °F. Water must be evaporated at the rate of 35,000 lb/hr. Neglecting boiling-point elevation, what must be the heating surface required? What is the steam consumption?

PROBLEM-SOLVING STRATEGIES

A. This is a conduction problem with resistances in series. It is also an optimization problem, that is, the minimum wall thickness is required. But it would be difficult to set up as a classical minimization problem. We shall use

$$q = \Delta T / \sum R_i$$

with $R_i = x_i / k_i A$. Start with $q = 250$ and ΔT across firebrick a minimum of $2000-1800°$ (to insure that insulating brick temperature is less than $1800°$ everywhere). This determines a minimum x_1. The actual x_1 will probably be larger since x_1 must be a multiple of 4.5 in. Continue in this way for each type of brick, in order. Finally, knowing all of the actual x_i, recalculate the q.

B. This looks like a natural convection problem. But, with most natural convection situations, radiation should also be considered. Natural convection problems use the Grashof number. Radiation problems require a shape-dependent equation.

C. This is an unsteady-state conduction problem. For simple shapes, analytical solutions are available, but they are series solutions that are time consuming. The analytical solutions have been plotted in most textbooks. The fact that

$$h \gg \frac{k}{r_m}$$

means that we can ignore convection from the surface.

D. This is a forced convection problem and we need a heat transfer coefficient. Since heat transfer varies with tube arrangement, we must be able to find a graph or correlation for this type of tube arrangement. If we can find it, we have an easy problem.

E. This is a forced convection problem. We have enough information given to calculate the q needed to heat the water. We must find a correlation for the heat transfer coefficient for flow inside the tube—a Colburn equation. Assuming no fouling, we can calculate U. Knowing U and q, with all temperatures known, we can calculate the heat transfer area (and thus the tube length) needed. If the overall heat transfer coefficient decreases with time on stream, it is the result of a fouling factor term in U. An easy problem.

F. This is a steady-state conduction problem with resistances in series. We can assume that the outside temperature of the 85 % magnesia is at atmospheric temperature and the inside pipe wall temperature is at the inside temperature. Because of the circular geometry, we must use log mean areas. This is a straightforward, simple problem.

G. This is a forced convection heat exchanger problem. We are given enough information to calculate the q required to heat the cold water. Then we can calculate the outlet hot-water temperature. Since we are given the U, we can calculate the heat exchanger area. All of the foregoing was for present operation. For the initial design conditions, the area is the same and the U is reduced by the dirt factor. So we know everything but the two outlet temperatures. We have enough equations to calculate these temperatures, but because of the log mean temperature, this would be a trial-and-error solution.

H. This is a standard evaporator problem requiring only heat and material balances. All the data needed are given and the solution should be straight-forward.

SOLUTIONS

A. See Figure H7. x_1, x_2, and x_3 must be multiples of 4.5, 3, and 4 in., respectively.

Basis: 1 ft² of surface area

We use the series formula

$$q = \Delta T / \sum R, \qquad R = x/kA, \qquad A = 1 \text{ ft}^2$$
$$q = 250(\text{Btu/hr ft}^2)(1 \text{ ft}^2) = 250 \text{ Btu/hr}$$

For firebrick (a), $\Delta T_1 \geq (2000\text{–}1800°F) = 200°F$

$$R = \frac{x_1}{k_1 A} = \frac{\Delta T_1}{q}, \qquad x_1 = k_1 A \frac{\Delta T_1}{q}$$

$$x_1 = 0.9 \frac{\text{Btu ft}}{\text{hr ft}^2 \, °F} (1 \text{ ft}^2) \frac{200°F}{250 \text{ Btu/hr}} = 0.72 \text{ ft} = 8.64 \text{ in.}$$

So we need two layers of firebrick ($x_1 = 2 \times 4.5$ in. $= 9$ in. $= 0.75$ ft). Then

$$\Delta T_1 = \frac{x_1}{k_1 A} q = 0.75 \text{ ft } 250 \text{ Btu/hr} \bigg/ \frac{0.9 \text{ Btu ft}}{\text{hr ft}^2 \, °F} (1 \text{ ft}^2) = 208°F$$

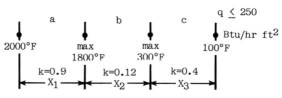

FIGURE H7. Schematic for solution A.

Then

$$\Delta T_2 \geq 2000°F - 208°F - 300°F = 1492°F$$

$$x_2 = 0.12 \frac{Btu\ ft}{hr\ ft°F}(1\ ft^2)\frac{1492°F}{250\ Btu/hr} = 0.716\ ft = 8.59\ in.$$

So we need three layers of insulating (b), $[x_2 = 3(3\ in.) = 9\ in. = 0.75\ ft]$.
Then

$$\Delta T_2 = \frac{x_2}{k_2 A}q = \frac{0.75(250)}{0.12(1)} = 1563°F$$

Now

$$\Delta T_3 \geq (2000°F - 208°F - 1563°F - 100°F) = 129°F$$

$$x_3 = 0.4(1)(129/250) = 0.206\ ft = 2.48\ in.$$

So we need one layer of building (c), $[x_3 = 1(4\ in.) = 0.33\ ft]$.
   Now minimum wall thickness = 9 in. + 9 in. + 4 in. = 22 in. and actual heat loss is

$$q = \Delta T / \sum R = (2000 - 100)°F \bigg/ \left(\frac{0.75}{0.9} + \frac{0.75}{0.12} + \frac{0.333}{0.4}\right)$$

$$= \frac{1900}{7.916} = 240\ \frac{Btu}{hr\ ft^2}$$

B. See Figure H8. This is a natural or free convection problem—and possibly a radiation problem. Gr is the Grashof number.
   For convection

$$q_c = hA\ \Delta T \qquad A = 10 \times 20\ ft^2 = 200\ ft^2$$
$$\Delta T = 80{-}0°F \quad = 80°F$$

$$\frac{Ub}{k_f} = 0.0317\ Gr_b^{0.37} \qquad \text{if } Gr_b > 2 \times 10^5;\ U = h \qquad \text{(Kreith, Eq. 7-36)}$$

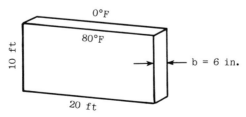

FIGURE H8. Schematic for solution B.

For radiation

$$q_r = \frac{\sigma A(T_1^4 - T_2^4)}{(1/\varepsilon_1 + 1/\varepsilon_2 - 1)} \quad \text{(McCabe and Smith, Eq. 14-39)}$$

$$\text{Gr}_b = \frac{\rho^2 g \beta \Delta T b^3}{\mu^2}, \quad \text{at} \sim 40°\text{F}, \quad \frac{q\beta\rho^2}{\mu^2} = 3.16 \times 10^6 \frac{1}{°\text{F ft}^3} \text{ (Kreith)},$$

$$k = 0.0140 \text{ Btu/hr ft°F}$$

$$\text{Gr}_b = (3.16 \times 10^6/°\text{F ft}^3)(80°\text{F})(\tfrac{1}{2} \text{ ft})^3 = 3.16 \times 10^7$$

$$U = h = \frac{0.0317(3.16 \times 10^7)^{0.37}}{(\tfrac{1}{2} \text{ ft})}\left(\frac{0.0140 \text{ Btu}}{\text{hr ft°F}}\right) = 0.5286 \frac{\text{Btu}}{\text{hr ft}^2 \, °\text{F}}$$

$$q_c = 0.5286 \text{ Btu/hr ft}^2 \, °\text{F} (200 \text{ ft}^2)(80°\text{F}) = 8457 \text{ Btu/hr}$$

$$\varepsilon_1 = \varepsilon_2 = 0.93, \text{ (Kreith Appendix)}, \qquad T_1 = 460 + 80 = 540°\text{R},$$

$$T_2 = 460 + 0 = 460°\text{R}$$

$$\sigma = 0.1714 \times 10^{-8} \frac{\text{Btu}}{\text{hr ft}^2 \, °\text{R}^4}$$

$$q_r = 0.1714 \times 10^{-8}(200)\left/\left(\frac{1}{0.93} + \frac{1}{0.93} - 1\right)(540^4 - 460^4) = 11{,}994 \frac{\text{Btu}}{\text{hr}}\right.$$

$$q = q_c + q_r = 20{,}451 \text{ Btu/hr from one wall}$$

For Al foil $\varepsilon_1 = 0.93$, $\varepsilon_2 = 0.04$, (Kreith), $b = \tfrac{1}{2}(6 \text{ in.}) = 3 \text{ in.} = \tfrac{1}{4} \text{ ft}$. Assume physical properties to be about the same; $\Delta T = \tfrac{1}{2}(80°\text{F}) = 40°\text{F}$.

$$\text{Gr}_b = (3.16 \times 10^6/°\text{F ft}^3)40°\text{F}(\tfrac{1}{4} \text{ ft})^3 = 1.975 \times 10^6$$

$$U = h = 4[0.0317(1.975 \times 10^6)^{0.37} \times 0.0140] = 0.3790$$

$$q_c = 0.3790(200)(40) = 3032 \text{ Btu/hr}, \qquad T_1 = 40 + 460 = 500°\text{R}$$

$$T_2 = 0 + 460 = 460°\text{R}$$

$$q_r = 0.1714 \times 10^{-8}(200)\left/\left(\frac{1}{0.93} + \frac{1}{0.04} - 1\right)[(500)^4 - (460)^4] = 242 \frac{\text{Btu}}{\text{hr}}\right.$$

$$q = q_c + q_r = 3032 + 242 = 3274 \text{ Btu/hr for one wall}$$

Use the new engineer's idea and give her a raise.

C. This is an unsteady-state conduction problem in an infinitely large slab. Kreith, p. 166, has charts for this type of problem.

We need the following parameters: thickness $= 2L = 4$ in., so $L = 2$ in.; $x/L = 0/2$ in. $= 0$ centerline; Biot modulus $hr_m/k_s \cong \infty$; Fourier modulus $a\theta/L^2$, $a = $ thermal diffusivity $= k/C_p\rho$.

$$\frac{T_{x/L} - T_\infty}{T_0 - T_\infty} = \frac{100°\text{F} - 50°\text{F}}{400°\text{F} - 50°\text{F}} = \frac{50}{350} = \frac{1}{7} = 0.143$$

$$\frac{a\theta}{L^2} = 0.9, \qquad \theta = 0.9\left(\frac{2 \text{ in.}}{12 \text{ in./ft}}\right)^2 \frac{\text{hr ft°F } 0.18 \text{ Btu } 160 \text{ lb}}{0.36 \text{ Btu lb°F ft}^3}$$

$$\theta = 2 \text{ hr}$$

D. $q = hA\,\Delta T_{\mathrm{lm}} = \dot{m}C_p(T_o - T_i)$, so to get T_o we must first find the heat transfer coefficient. Kreith has a section on tube banks, pp. 415–425. Pr is the Prandtl number.

From his equations,

$$A_{\min}\left(\frac{\text{min free area}}{\text{unit length}}\right) = S_T - D_o$$

where S_T = distance between centers.

Basis: 1-ft length of bundle

$$A_{\min} = \frac{1.5\ \text{in.} - 0.75\ \text{in.}}{12\ \text{in./ft}} = 0.0625\ \text{ft}^2/\text{ft}, \qquad \begin{aligned} D_o &= \tfrac{3}{4}/12 = 0.0625\ \text{ft}\\ A_o &= 0.1963\ \text{ft}^2/\text{ft}\end{aligned}$$

and

$$\frac{hD_o}{k_f} = 0.33\left(\frac{G_{\max}D_o}{\mu_f}\right)^{0.6}\mathrm{Pr}_f^{0.3} \quad \text{if} \quad \frac{G_{\max}D_o}{\mu_f} \ge 6000, \qquad \text{(Kreith, Eq. 9-10)}$$

where f says use mean film temperature, that is,

$$\tfrac{1}{2}[\tfrac{1}{2}(T_o + T_i) + T_w]$$

$$G_{\max} = \frac{5000\ \text{lb}}{\text{hr}}\frac{\text{hr}}{\text{ft}^2}\frac{}{3600\ \text{sec}} = 1.389\ \frac{\text{lb}}{\text{ft}^2\ \text{sec}}$$

Assume $T_o = 100°\text{F}$, then

$$T_f = \tfrac{1}{2}[\tfrac{1}{2}(100 + 50) + 200] = 137.5°\text{F}$$

From Kreith, Appendix A-3,

$$\mathrm{Pr} = 0.72, \qquad \mu_f = 1.34 \times 10^{-5}\ \text{lb/ft sec},$$

$$k = 0.0162\ \text{Btu/hr ft}°\text{F}, \qquad C_p(75°\text{F}) = 0.24\ \text{Btu/lb}°\text{F}$$

$$\frac{G_{\max}D_o}{\mu_f} = 1.389\ \frac{\text{lb}}{\text{ft}^2\ \text{sec}}\frac{(0.0625\ \text{ft})\ \text{ft sec}}{1.34 \times 10^{-5}\ \text{lb}} = 6479$$

$$h = 0.33(6479)^{0.6}(0.72)^{0.3}\left(\frac{0.0162\ \text{Btu}}{\text{hr ft}°\text{F}}\right)\frac{1}{0.0625\ \text{ft}} = 15.0\ \frac{\text{Btu}}{\text{hr ft}^2\ °\text{F}}$$

$$A_s = 144\left(0.1963\ \frac{\text{ft}^2}{\text{ft}}\right) = 28.67\ \text{ft}^2, \qquad \dot{m} = \frac{5000\ \text{lb}}{\text{hr ft}}(0.0625\ \text{ft}^2)12 = 3750\ \frac{\text{lb}}{\text{hr}}$$

$$\Delta T_{\mathrm{lm}} = \frac{(200 - 50) - (200 - 100)}{\ln\,[(200 - 50)/(200 - 10)]} = \frac{150 - 100}{\ln\,(150/100)} = 123.3°\text{F}$$

So

$$q = 15.0\ \frac{\text{Btu}}{\text{hr ft}^2\ °\text{F}}(28.67\ \text{ft}^2)(123.3°\text{F}) = 3750\ \frac{\text{lb}}{\text{lb}°\text{F}}\left(0.24\ \frac{\text{Btu}}{\text{lb}°\text{F}}\right)(T_o - 50)°\text{F}$$

or

$$T_o = 58.9 + 50 = 108.9°F$$

Second approximation:

$$T_o = 108°F, \qquad T_f = \tfrac{1}{2}[\tfrac{1}{2}(108 + 50) + 200] = 140°F$$

So physical properties remain the same

$$T_{lm} = \frac{(200 - 50) - (200 - 108)}{\ln\left[(200 - 50)/(200 - 108)\right]} = \frac{150 - 92}{\ln(150/92)} = 118.7°F$$

$$q = 15(28.67)(118.9) = 3750(0.24)(T_o - 50)$$

or

$$T_o = 56.8 + 50 = 106.8, \qquad T_o = 107°F$$

E.
$$\frac{1}{U_i} = \frac{1}{h_{di}} + \frac{1}{h_i} + \frac{x_w D_i}{k_w D} + \frac{1}{h_o}\frac{D_i}{D_o}$$

Initially, $h_{di} = 0$; 16 BWG, (Kreith, Appendix Table A5), Nu is the Nusselt number,

$$D_o = \tfrac{5}{8} \text{ in.}, \qquad D_i = 0.495 \text{ in.}, \qquad x_w = 0.065 \text{ in.}$$

$$A_{so} = 0.1636 \text{ ft}^2/\text{ft}, \qquad \begin{array}{l} A_{cs} = 0.0013 \text{ ft}^2, \\ A_{si} = 0.1296 \text{ ft}^2/\text{ft} \end{array}$$

$$\bar{D}_{lm} = \log \text{ mean } D, \qquad k_w = 26 \, \frac{\text{Btu}}{\text{hr ft}°F} \text{ assumed steel}$$

$$\bar{D}_{lm} = \frac{\tfrac{5}{8} - 0.495}{\ln(\tfrac{5}{8}/0.495)} = 0.557 \text{ in.}, \qquad \text{Nu} = 0.023 \, \text{Re}^{0.8}\text{Pr}^{1/3}$$

Evaluate H_2O properties at $T = (60 + 180°F)/2 = 120°F$ (Kreith, Appendix Table A3),

$$\text{Pr} = 3.81, \qquad k = 0.372 \, \frac{\text{Btu}}{\text{hr ft}°F}, \qquad C_p = 1 \, \frac{\text{Btu}}{\text{lb}°F},$$

$$\mu = 0.392 \times 10^{-3} \text{ lb/ft sec}$$

$$\text{Re} = \frac{GD}{\mu}, \qquad G = \frac{45 \text{ lb}}{\min(0.0013\text{ft}^2)} \frac{1 \text{ sec}}{60 \text{ min}} = 577 \, \frac{\text{lb}}{\text{ft}^2 \text{ sec}} \text{ or lb/ft}^2 \text{ sec}$$

$$\text{Re} = \frac{577 \text{ lb}}{\text{ft}^2 \text{ sec}} \frac{\text{ft sec}}{0.392 \times 10^{-3} \text{ lb}} \frac{0.495 \text{ in.}}{12 \text{ in./ft}} = 60,717$$

$$\text{Nu} = \frac{h_i D_i}{k} = 0.023(60,717)^{0.8}(3.81)^{1/3} = 241$$

$$h_i = 241 \frac{k}{D_i} = 241 \frac{0.372 \text{ Btu}}{\text{hr ft}°F} \frac{12 \text{ in./ft}}{0.495 \text{ in.}} = 2173 \frac{\text{Btu}}{\text{hr ft}^3 \,°F}$$

$$\frac{1}{U_i} = \frac{1}{2173} + \left(\frac{0.065}{26.0}\right)\left(\frac{1}{12}\right)\left(\frac{0.495}{0.557}\right) + \left(\frac{1}{1000}\right)(\tfrac{8}{5})(0.495)$$

$$= 1.437 \times 10^{-3}$$

$$U_i = 696 \text{ Btu/hr ft}^2 \,°F$$

$$q = U_i A_i \Delta T_{\text{lm}} = \dot{m} C_p (T_o - T_i);$$

$$\Delta T_{\text{lm}} = [(260 - 60) - (260 - 180)] \Big/ \ln\left(\frac{(260 - 60)}{(260 - 180)}\right) = 131°F$$

$$\dot{m} = \frac{45 \text{ lb}}{\text{min}},$$

$$q = \left(45 \frac{\text{lb}}{\text{min}}\right)\left(\frac{\text{Btu}}{\text{lb}°F}\right)(180 - 60)°F$$

$$= 696 \frac{\text{Btu}}{\text{hr ft}^2 \,°F}\left(0.1296 \frac{\text{ft}^2}{\text{ft}}\right) \frac{L \text{ ft } 131°F}{60 \text{ min/hr}}$$

So $L = 27.4$ ft. With scale

$$U_i = \tfrac{1}{2}(696) = 348, \qquad \frac{1}{h_{\text{di}}} + 1.437 \times 10^{-3} = \frac{1}{348},$$

$$\frac{1}{h_{\text{di}}} = 1.437 \times 10^{-3} \qquad \text{or} \qquad h_{\text{di}} = 696 \text{ Btu/hr ft}^2 \,°F$$

F. Kreith, Appendix Table A6,

$$D_o = 1.660 \text{ in.}, \qquad D_i = 1.380 \text{ in.}, \qquad x_w = 0.140 \text{ in.}$$

For resistances in series,

$$q = \Delta T / \sum R$$

For hollow cylinders,

$$R = \frac{r_o - r_i}{k\bar{A}}, \qquad \bar{A} = \frac{A_o - A_i}{\ln (A_o / A_i)}$$

Basis: 1 ft of length

then $A = \Pi D (1 \text{ ft}) = \Pi D$

pipe wall $R_w = \dfrac{x_w}{k_s A_w}$, $\bar{A}_w = \Pi \dfrac{D_o - D_i}{\ln (D_o / D_i)} = 0.3968 \text{ ft}^2$ wall $\begin{array}{l} D_i = 1.380 \text{ in.} \\ D_o = 1.660 \text{ in.} \end{array}$

$$R_w = 1.1308 \times 10^{-3}$$

asbestos $R_a = \dfrac{\frac{2}{12}\text{ft}}{0.087\bar{A}_a}$,

$\bar{A}_a = \Pi\dfrac{(D_o - D_i)_a}{\ln(D_o/D_i)} = 0.8537\text{ ft}^2$ asbestos $\begin{array}{l}D_i = 1.660\text{ in.}\\ D_o = 5.660\text{ in.}\end{array}$

$R_a = 2.244$

magnesia $R_m = \dfrac{\frac{3}{12}\text{ft}}{0.034A_m}$,

$\bar{A}_m = \Pi\dfrac{(D_o - D_i)_m}{\ln(D_o/D_i)} = 2.1734\text{ ft}^2$ magnesia $\begin{array}{l}D_i = 5.660\text{ in.}\\ D_o = 11.660\text{ in.}\end{array}$

$R_m = 3.3832$

$$q = \frac{300 - 80}{1.1308 \times 10^{-3} + 2.2440 + 3.3832}$$

$$= \frac{220}{5.6283}$$

$$= 39.1 \frac{\text{Btu}}{\text{per hr}} \text{ per ft of length}$$

$$39.1 = \frac{300 - T}{1.1308 \times 10^{-3} + 2.244} = \frac{300 - T}{2.2451}, \qquad T = 212.2°\text{F}$$

G. See Figure H9. We need the F_G factor from McCabe–Smith, Fig. 15.6a

$$Z = \frac{T_{ha} - T_{hb}}{T_{cb} - T_{ca}}, \qquad \eta = \frac{T_{cb} - T_{ca}}{T_{ha} - T_{ca}}$$

$$Z = \frac{\dot{m}_c C_{pc}}{\dot{m}_H C_{pH}}$$

$$q = U_o A_o F_G \Delta T_{lm} = m_H C_{ph}(T_{ha} - T_{hb}) = m_c C_{pc}(T_{cb} - T_{ca})$$

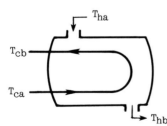

FIGURE H9.   Schematic for solution G.

To get C_p values, we need an estimate of T_{hb}:

$$\dot{m}_H(240 - T_{hb}) = \dot{m}_c(130 - 60)$$

$$T_{hb} = 222°F, \qquad T_h = 230°F, \qquad T_c = 95°F$$

Kreith, Appendix Table A3

$$C_{ph} = 1.005, \qquad C_{pc} = 0.9975 \text{ Btu/lb°F},$$

$$\rho_{240} = 58.9 \text{ lb/ft}^3, \quad \rho_{60} = 62.3 \text{ lb/ft}^3$$

$$\dot{m}_c = \frac{55 \text{ gal}}{\text{min}} \frac{1 \text{ ft}^3}{7.48 \text{ gal}} \frac{62.3 \text{ lb}}{\text{ft}^3} = 458.1 \frac{\text{lb}}{\text{min}}$$

$$= 2.749 \times 10^4 \text{ lb/hr}$$

$$\dot{m}_H = \frac{210 \text{ gal}}{\text{min}} \frac{1 \text{ ft}^3}{7.48 \text{ gal}} \frac{58.9 \text{ lb}}{\text{ft}^3} = 1654 \frac{\text{lb}}{\text{min}}$$

$$= 9.924 \times 10^4 \text{ lb/hr}$$

$$q_c = 2.749 \times 10^4 (0.9975)(70) = 1.919 \times 10^6 \text{ Btu/hr};$$

$$q_h = 9.924 \times 10^4 (1.005)(240 - T_{hb})$$

$$240 - T_{hb} = 19.24, \qquad T_{hb} = 220.8°F$$

$$\Delta \bar{T}_{lm} = [(220.8 - 60) - (240 - 130)]/\ln [(220.8 - 60)/(240 - 130)] = 133.8$$

$$Z = 0.274, \qquad F_G = 0.99, \qquad \eta = 0.389$$

So our exchanger has an A_o of

$$\frac{q}{U_o F_G \Delta T_{lm}} = \frac{1.919 \times 10^6}{300(0.99)(133.8)} = 48.3 \text{ ft}^2$$

Now from Kreith, p. 505

$$R_d = 0.002 = \frac{1}{(U_o)_{dirty}} - \frac{1}{(U_o)_{clean}} = \frac{1}{300} - \frac{1}{(U_o)_{clean}}$$

So

$$(U_o)_{clean} = 750 \text{ Btu/hr ft}^2 \text{ °F}$$

We need to find T_{hb} and T_{cb} with this new U_o

(1) $q = U_o A_o F_G \Delta \bar{T}_{lm} = 750(48.3)(0.99)\Delta \bar{T}_{lm} = 35863 \Delta \bar{T}_{lm}$

(2) $q_c = \dot{m}_c C_{pc}(T_{cb} - 60) = 2.749 \times 10^4 (0.9975)(T_{cb} - 60)$
$= 27421(T_{cb} - 60)$

(3) $q_h = \dot{m}_H C_{pH}(240 - T_{hb}) = 9.924 \times 10^4(1.005)(240 - T_{hb})$
     $= 99736(240 - T_{hb})$

$$\Delta \bar{T}_{lm} = [(T_{hb} - 60) - (240 - T_{cb})] \left/ \ln \left(\frac{T_{hb} - 60}{240 - T_{cb}} \right) \right.$$

(A)
(2) = (3) $T_{hb} = (256.5 - 0.2749 T_{cb})$

(B)
(1) = (2) $T_{cb} = 1.3079 \left[\dfrac{\overset{a}{\overbrace{(196.5 - 0.2749 T_{cb})}} - \overset{b}{\overbrace{(240 - T_{cb})}}}{\ln (a/b)} \right] + 60$

(C)	(A) T_{hb}	a	b	(B)	
$T_{cb}(°F)$	$(256.5-0.2749 T_{cb})$	$(196.5-0.2749 T_{cb})$	$(240 - T_{cb})$	T_{cb}	(B) − (C)
140	218.0	158.0	100	225.8	85.8
160	212.5	152.5	80	207.0	47.0
180	207.0	147.0	60	187.0	+7.0
184	205.9	145.9	50	182.8	−1.2
183	206.2	146.2	57	183.9	+0.9
183.5	206.1	146.1	56.5	183.3	+0.2

H. See Figure H10.

$$Q + \dot{m}_f C_p(T_f - T) = (\dot{m}_f - \dot{m})H_V + \dot{m} C_p(T - \bar{T})$$

$$\bar{T} = 32, \qquad \dot{m}_f - \dot{m} = 35,000 \text{ lb/hr}$$

$$C_p = 0.93 \text{ Btu/lb°F}$$

$$T_f = 50°F, \qquad T = 133.8°F$$

Enthalpy balance:

$$Q + 0.93(50 - 32)\dot{m}_f = 35,000 \times 1119.4 + 0.93(133.8 - 32)\dot{m}$$

Solids balance:

$$0.15 \dot{m}_f = 0.65 \dot{m}, \qquad \dot{m}_f = (0.65/0.15)\dot{m}$$

Total balance:

$$\dot{m}_f = 35,000 + \dot{m} \qquad \text{or} \qquad \frac{0.65}{0.15} \dot{m} = 35,000 + \dot{m}$$

$$\frac{0.65 - 0.15}{0.15} \dot{m} = 35,000, \qquad \dot{m} = \frac{35,000}{3.333} = 10,500$$

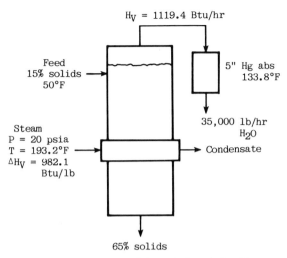

FIGURE H10. Schematic for solution H.

Then

$$\dot{m}_f = 45,500$$

$$Q = 35,000 \times 1119.4 + 0.93(133.8 - 32)10,500 - 0.93(50 - 32)45,500$$
$$= 39,411,407 \text{ Btu/hr}$$

$$\Delta t = 193.2°F - 133.8°F = 59.4°F$$

$$A = \frac{Q}{U\Delta t} = \frac{39,411,407}{200 \times 59.4} = 3317 \text{ ft}^2$$

$$\text{steam required} = \frac{39,411,407 \text{ Btu/hr}}{982.1 \text{ Btu/lb}} = 40,130 \text{ lb/hr}$$

CHEMICAL KINETICS

O. Levenspiel, *Chemical Reaction Engineering*, 2nd ed., Wiley, New York, 1972.

I. INTRODUCTION

We shall discuss the basic principles of kinetics and reactors. Only topics likely to be of use on the test will be covered. The reference on p. 173 (Carberry or Smith texts in McGraw-Hill's Chemical Engineering Series are also good) should be consulted for points not discussed here. Remember that there will probably be more kinetics problems than any other type.

II. THE BASICS OF KINETICS

A. Definition of Reaction Rate

In homogeneous reactions all reacting materials are found within a single phase: gas, liquid, or solid. In addition, if the reaction is catalytic, the catalyst must also be present within the same phase as the reactants and products. Although there are a number of ways of defining the rate of reaction, only one such measure is used in homogeneous systems. This is the intensive measure of rate based on unit volume of reacting fluid or system.

Consider a reaction in which substances A and B react to form C. (It is irrelevant and confusing to put down the chemical equation at this point.) Let N_a, N_b, and N_c be the number of moles of these substances present at any moment in a system of fixed mass and of volume V. If the system is uniform in composition — as we shall suppose — the concentrations of the various substances are

$$[A] = N_a/V; \quad [B] = N_b/V; \quad [C] = N_c/V$$

The reaction rate r is the rate, per unit volume of the reacting fluid, at which any one of the substances, chosen as a key substance, is being produced or destroyed by reaction. Let it be the product C which is so chosen. Then if — and only if — there is no process involving C other than the reaction, the rate r may be written

$$r = \frac{1}{V}\frac{dN_c}{dt} = \frac{1}{V}\frac{d(V[C])}{dt}$$

This equation would not be applicable if substance C were being continually introduced into or removed from the given volume of a physical process. The equation as it stands applies only to closed systems and it does not apply, for example, to a CFSTR (continuous flow stirred tank reactor). In the latter the substance C is being continually removed through the outflow and dN_c/dt (but not r) is zero at the steady state of the system.

It would be expected that the variables that affect the progress of homogeneous reactions are the composition of the materials within the phase as well as the temperature and pressure of the system. The container shape, surface properties of solid materials in contact with the phase in question, and the diffusional characteristics of the fluid should not affect the rate of homogeneous reaction for component C:

$$r = f(\text{temperature, pressure, composition})$$

The variables temperature, pressure, and composition are interdependent, since the pressure is fixed given the temperature and composition of the phase. So we can write without loss of generality

$$r = f(\text{temperature, composition})$$

B. Single and Multiple Reactions

Before finding the form of the concentration term in a rate expression, we shall distinguish between a number of types of reactions. This distinction is based on the form and number of kinetic equations used to describe the process of a chemical reaction. Since we are concerned with the concentration-dependent term of rate equations for the time being, we assume in the following discussion that the temperature of the system is kept constant.

When materials react to form products it is usually easy to decide after examining the stoichiometry, preferably at more than one temperature, whether we should consider a single reaction or a number of reactions to be occurring. When a single stoichiometric equation and single rate equation are chosen to represent the progress of the reaction, we have a single reaction. When more than one kinetic expression is needed to follow the changes in composition of all components of the reaction we have multiple reactions.

Multiple reactions are classified as consecutive or series reactions,

$$A \rightarrow R \rightarrow S$$

parallel, competing, or side reactions,

$$
\begin{array}{ccc}
A \rightarrow R & & A \rightarrow R \\
& \text{or} & \\
A \rightarrow S & & B \rightarrow S
\end{array}
$$

and mixed reactions

$$A + B \rightarrow R$$
$$R + B \rightarrow S$$

The mixed reactions illustrated are parallel with respect to B and consecutive with respect to A, R, and S.

C. Molecularity and Order

The molecularity of an elementary reaction is the number of molecules involved in the rate-determining step of a reaction. Molecularity of reactions has been found to be one, two, and occasionally three. Needless to say, the molecularity refers only to an elementary reaction and can only be in whole numbers.

Often we find that the rate of progress of a reaction, involving materials A, B, ..., D, can be approximated by an expression of the following type:

$$r_a = kC_a^a C_b^b \cdots C_d^d, \qquad a + b + \cdots + d = n$$

where the a, b, ... are not necessarily related to the stoichiometric coefficients. We call the powers to which the concentrations are raised the order of the reaction. Thus the reaction is

ath order with respect to A

bth order with respect to B

nth order overall

Since the order refers to the empirically found rate expression, they need not be whole numbers, but the molecularity of a reaction must be in terms of whole numbers since it refers to the actual mechanism of the elementary reaction.

D. Rate Constant k

When a rate expression for a homogeneous chemical reaction is written in the form shown in the previous equation, the units of the rate constant k for the nth-order reaction are

$$(\text{time})^{-1}(\text{concentration})^{1-n}$$

which becomes, for a first-order reaction, simply

$$(\text{time})^{-1}$$

E. Representation of a Reaction Rate

In writing a rate equation we may use any measures equivalent to concentration, such as the partial pressures of the components. So

$$r_A = k p_a^a p_b^b \cdots p_d^d$$

Whatever measure is used leaves the order unchanged; but it will affect the units of the rate constant k.

For shortness, elementary reactions are often represented by an equation showing both the molecularity and the rate constant. Thus

$$2A \xrightarrow{k_1} 2R$$

represents a bimolecular irreversible reaction with second-order rate constant k_1 implying that the rate equation is

$$-r_a = r_r = k_1 C_a^2$$

It would not be proper to write the stoichiometric equation for this reaction as

$$A \xrightarrow{k_1} R$$

for this would imply that the rate expression is

$$-r_a = r_r = k_1 C_a$$

We must be careful to distinguish between a stoichiometric equation that can be multiplied by any constant and the equation that represents an elementary reaction.

Even writing the equation correctly may not be sufficient to avoid ambiguity for some reactions. For instance, consider the reaction

$$B + 2D \overset{k_2}{\to} 3T$$

If the rate were measured in terms of B, the rate equation is

$$-r_b = k_2' C_b C_d^2$$

But if it refers to D, the rate equation is

$$-r_d = k_2'' C_b C_d^2$$

Or if it refers to the product T,

$$r_t = k_2''' C_b C_d^2$$

But from the stoichiometry

$$r_b = \tfrac{1}{2} r_d = -\tfrac{1}{3} r_t$$

so

$$k_2' = \tfrac{1}{2} k_2'' = \tfrac{1}{3} k_2'''$$

To which of these primed k_2 values are we referring in the stoichiometric equation? We must be told.

There is an easy way around this problem, but the method is not widely used. First we set the stoichiometric equation up with all quantities on one side, the product stoichiometric coefficients being positive, those for the reactants negative. The previous stoichiometric equation would be written

$$3T - 1B - 2D = 0$$

The rate would be defined as

$$r = \frac{1}{\alpha_i V} \frac{dN_i}{dt}$$

where the α_i is the stoichiometric coefficient of the ith component. Thus the rate would always be a positive quantity. The rate expression corresponding to this would become

$$r = k_2 C_b C_d^2 = \frac{1}{3V} \frac{dN_t}{dt} = -\frac{1}{V} \frac{dN_b}{dt} = -\frac{1}{2V} \frac{dN_d}{dt}$$

F.   Material Balance in Complex Reactions

In simple cases, the relations between the conversions of the several participants can be found easily by inspection, but in complex ones a systematic procedure is helpful. The reactions of a group can be numbered, and the initial and final amounts of the participants identified for each reaction. For those substances that participate in more than one reaction, the initial amount for a particular reaction is the same as the final amount from the last preceding reaction in which this substance participated. It is convenient to subscript the final amount from any reaction with the number of that reaction. Then all the possible material balances are written down, between initial, intermediate, and final amounts. The balances may then be combined and rearranged to eliminate the intermediate amounts, leaving only the initial and final amounts for the complete reaction.

For example,

Reaction 1:	$A + B \rightarrow C + E$
Reaction 2:	$A + C \rightarrow D + E$
Reaction 1, initial amounts:	A_0, B_0, C_0, E_0 (no D involved)
Reaction 1, final amounts:	A_1, B, C_1, E_1
Reaction 2, initial amounts:	A_1, C_1, D_0, E_1 (no B involved)
Reaction 2, final amounts:	A, C, D, E
Reaction 1, balance:	$A_0 - A_1 = B_0 - B = C_1 - C_0 = E_1 - E_0$
Reaction 2, balance:	$A_1 - A = C_1 - C = D - D_0 = E - E_1$

Using algebraic elimination of the quantities with subscript 1, the material balance is expressed in these three relations:

$$A_0 - 2B_0 - C_0 = A - 2B - C$$

$$B_0 + C_0 + D_0 = B + C + D$$

$$A_0 + E_0 = A + E$$

As another example, consider

Reaction 1:	$2A \rightarrow B$
Reaction 2:	$2B \rightarrow C$
Reaction 1, initial:	A_0, B_0
Reaction 1, final:	A, B_1
Reaction 2, initial:	B_1, C_0
Reaction 2, final:	B, C
Reaction 1, balance:	$\frac{1}{2}(A_0 - A) = B_1 - B_0$
Reaction 2, balance:	$\frac{1}{2}(B_1 - B) = C - C_0$

Upon eliminating B_1, the final balance is

$$A_0 - A + 2(B_0 - B) = 4(C - C_0)$$

G.   Temperature Dependency from the Arrhenius Law

For many reactions, and especially elementary reactions, the rate expression may be written as a product of a temperature-dependent term and a composition-dependent term, or

$$r_i = f_1(\text{temperature}) \cdot f_2(\text{composition})$$
$$= k \cdot f_2(\text{composition})$$

For these reactions the temperature-dependent term k, the reaction rate constant (velocity constant), has been found in most cases to be well represented by Arrhenius' law:

$$k = k_0 \exp\left(-\frac{E}{RT}\right)$$

where k_0 is called the frequency factor and E is called the activation energy of the reaction. This equation fits experimental data quite well over wide temperature ranges and is a first approximation to the temperature dependency derived according to several different theories.

III.   INTERPRETATION OF BATCH REACTOR DATA

A.   Introduction

A rate equation characterizes the rate of reaction. The value of the constants of the equation can only be found by experiment. The determination of this rate equation is usually a two-step procedure: First the concentration dependency is found at fixed temperature, and then the temperature dependence of the rate constants are found.

The equipment used to obtain rate data are of two different types: the batch and the flow reactor. The batch reactor is simply a container to hold the contents while they react. An experimental batch reactor is usually operated isothermally and at constant volume because of the ease of interpretation of the results of such runs.

There are two procedures for analyzing experimental kinetic data, the integral and the differential methods. In the integral method, we select a kinetic model with its corresponding rate equation and, after appropriate mathematical integrations and manipulations, predict that a plot of C versus t data on specific x versus y coordinates should yield a straight line. The data are plotted and if a good straight line results, we accept the assumed kinetic model.

For the differential method of analysis we select a kinetic model and fit its corresponding rate expression to the data directly. But since the rate expression is a differential equation, we must first find $(1/V)(dN/dt)$ from the data before the fitting procedure is attempted.

The integral method is easy to use when fitting relatively simple mechanisms or when data are so scattered that we cannot reliably find the derivatives needed for the differential method. The differential method is more useful for the more complicated mechanisms but it requires accurate data, as well as more data.

B.   Constant-Volume Batch Reactor

The constant volume refers to the volume of the reactor taken up by the reaction mixture. This is thus a constant-density reactor. Most liquid-phase reactions as well as all gas-phase reactions occurring in a constant-volume bomb fall into this category.

With a constant-volume reactor, the measure of the reaction rate becomes

$$r_i = \frac{1}{V}\frac{dN_i}{dt} = \frac{d(C_i V)}{dt} = \frac{1}{V}\frac{C_i\,dV + V\,dC_i}{dt} = \frac{dC_i}{dt}$$

or for ideal gases

$$r_i = \frac{1}{RT}\frac{dp_i}{dt}$$

These measures of rate can be followed directly in most systems. Because of the ease of interpretation of such data, the constant volume system is used whenever possible, even though the plant built for the commercial application of the reaction will probably be a constant-pressure system.

1.   Irreversible Unimolecular-Type First-Order Reaction

Suppose for the reaction

$$A \rightarrow \text{products}$$

we wish to test the first-order rate equation

$$-\frac{dC_a}{dt} = kC_a$$

Separating and integrating we obtain

$$-\ln\frac{C_a}{C_{a0}} = kt$$

Now, by definition, the fractional conversion X_a of a given reactant is defined as the fraction of the reactant converted into product or

$$N_a = N_{a0}(1 - X_a)$$

where N_{a0} is the starting moles of A. Fractional conversion (or just conversion) is often used in place of concentration in engineering work; thus, most of the following results will be given in terms of both C_a and X_a. Now

$$C_a = \frac{N_a}{V} = \frac{N_{a0}(1 - X_a)}{V} = C_{a0}(1 - X_a)$$

So in terms of conversion, the first-order reaction equation becomes

$$-\ln(1 - X_a) = kt$$

2. Other Reaction Types

Equations for other orders of constant-volume reactions are shown in Table K1.

C. Variable-Volume Batch Reactor

The general form for the rate of change of component i in either the constant-or variable-volume reaction system has already been given as

$$r_i = \frac{1}{V}\frac{dN_i}{dt} = \frac{1}{V}\frac{d(C_i V)}{dt} = \frac{1}{V}\frac{V \, dC_i + C_i \, dV}{dt}$$

or

$$r_i = \frac{dC_i}{dt} + \frac{C_i}{V}\frac{dV}{dt}$$

We see that two terms must be evaluated from experiment if r_i is to be determined. Fortunately, for the constant-volume system the second term dropped out, leaving only the simple expression

$$r_i = \frac{dC_i}{dt}$$

In a variable-volume reactor, we can also avoid the use of the unwieldy two term expression if we use fractional conversion instead of concentration and if we are willing to assume that the volume of the reacting system varies linearly with conversion, or

$$V = V_0(1 + \varepsilon_a X_a)$$

where ε_a is the fractional change in the volume of the system between no conversion and complete conversion. Then

$$\varepsilon_a = \frac{V(X_a = 1) - V(X_a = 0)}{V(X_a = 0)}$$

TABLE K1.   Some Rate Equations

Reaction	Order	Rate	Concentration/Time Equations[a]
$A \rightarrow B$	0	$-dC_a/dt = k$	$C_a = C_{a0} - kt$; $T = C_{a0}/2k$
$A \rightarrow B$	1	$-dC_a/dt = kC_a$	$\ln(C_a/C_{a0}) = -kt$; $T = (\ln 2)/k$
$A + A \rightarrow P$	2	$-dC_a/dt = kC_a^2$	$(1/C_a) - (1/C_{a0}) = kt$; $T = 1/(k\,C_{a0})$
$aA + bB$ $\rightarrow P$	2	$-dC_a/dt = kC_aC_b$	$\ln(C_a/C_b) - \ln(C_{a0}/C_{b0})$ $= [(bC_{a0} - aC_{b0})/a]kt$; $T = a/k(bC_{a0} - aC_{b0}) \times$ $\ln[aC_{b0}/(2aC_{b0} - bC_{a0})]$
$A \rightleftarrows B$	1	$-dC_a/dt = kC_a - k'C_b$	$\ln[(C_{a0} - A')/(C_a - A')] = \bar{k}t$; $\bar{k} = k(K + 1)/K$; $K = k/k'$; $A' = (C_{b0} + C_{a0})/(K + 1)$
$A + B \rightleftarrows$ $C + D$	2	$-dC_a/dt = kC_aC_b - k'C_cC_d$	$\ln\left\{\dfrac{1 + 2cF/(b - \sqrt{f})}{1 + 2cF/(b + \sqrt{f})}\right\} = \sqrt{f}\,t$; $a = k[C_{a0}C_{b0} - C_{c0}C_{d0}/K]$; $c = k\,(1 - 1/K)$; $b = -k[(C_{a0} + C_{b0})$ $+ (C_{c0} + C_{d0})/K]$ $f = b^2 - 4ac$; $F = C_{a0} - C_a$

[a] T is the half-life for the reactions.

To illustrate this point, consider the isothermal gas-phase reaction

$$A \rightarrow 4R$$

if we start with pure A,

$$\varepsilon_a = \frac{4 - 1}{1} = 3$$

If we had started with 50% inerts present, two volumes of reactant mixture (1 pure A, 1 inerts) yield upon complete conversion five volumes of product mixture (4 R, 1 inerts), so

$$\varepsilon_a = \frac{5 - 2}{2} = 1.5$$

We thus note that ε_a accounts for both the reaction stoichiometry and the presence of inerts.

Now since

$$N_a = N_{a0}(1 - X_a)$$

We obtain

$$C_a = \frac{N_a}{V} = \frac{N_{a0}(1 - X_a)}{V_0(1 + \varepsilon_a X_a)} = C_{a0}\frac{1 - X_a}{1 + \varepsilon_a X_a}$$

or

$$\frac{C_a}{C_{a0}} = \frac{1 - X_a}{1 + \varepsilon_a X_a}$$

This is the conversion–concentration relationship that we shall use if we assume linear volume change. This assumption is quite good for isothermal constant-pressure systems (if there are no series reactions). The rate expression with the linearity assumption becomes

$$-r_a = -\frac{1}{V}\frac{dN_a}{dt} = -\frac{1}{V_0(1 + \varepsilon_a X_a)}\frac{N_{a0}\, d(1 - X_a)}{dt}$$

$$= \frac{C_{a0}}{1 + \varepsilon_a X_a}\frac{dX_a}{dt}$$

1. Integral Method of Analysis

The integral method of data analysis of isothermal variable-volume reactors requires that we integrate the rate expression we wish to test. Then the C versus t function obtained is compared with our experimental data. So for reactant A

$$-r_a = -\frac{1}{V}\frac{dN_a}{dt} = \frac{C_{a0}}{1 + \varepsilon_a X_a}\frac{dX_a}{dt}$$

or, upon formal integration

$$t = C_{a0}\int_0^{X_a}\frac{dX_a}{(1 + \varepsilon_a X_a)(-r_a)}$$

This is the expression to be used for all batch reactors for which we can assume that the volume varies linearly with conversion.

IV. SINGLE IDEAL REACTORS

A. Single Ideal Batch Reactors

We make a material balance for any component A. (We usually select the limiting component.) In a batch reactor, since the composition is assumed uniform, we make the balance around the whole reactor. Thus we have

production of A = accumulation of A

since there is no input or output. Now

$$\text{production of A} = r_a V \text{(moles A formed/time)}$$

$$\text{accumulation of A} = \frac{dN_a}{dt} = \frac{d}{dt}[N_{a0}(1 - X_a)] = -N_{a0}\frac{dX_a}{dt}$$

so

$$r_a V = -N_{a0}\frac{dX_a}{dt}$$

or, as it is usually written

$$(-r_a)V = N_{a0}\frac{dX_a}{dt}$$

which integrates to

$$t = N_{a0}\int_0^{X_a} \frac{dX_a}{(-r_a)V}$$

This is the general equation giving the time required to get a given conversion of reactant A for either isothermal or nonisothermal operation. We leave both V and $(-r_a)$ under the integral since, in general, they both vary as the reaction proceeds.

Of course, for a number of situations the integral simplifies. If the volume of the mixture remains constant we have

$$t = C_{a0}\int_0^{X_a} \frac{dX_a}{(-r_a)} = -\int_{C_{a0}}^{C_a} \frac{dC_a}{(-r_a)}$$

For all reactions in which the volume of the mixture changes linearly with conversion, we have

$$t = N_{a0}\int_0^{X_a} \frac{dX_a}{(-r_a)V_0(1 + \varepsilon_a X_a)} = C_{a0}\int_0^{X_a} \frac{dX_a}{(-r_a)(1 + \varepsilon_a X_a)}$$

B. Steady-State Backmix Flow Reactor (CFSTR)

Since the composition is uniform throughout the tank (by definition), we make the material balance over the whole tank. So, if we consider reactant A,

$$\text{input-output} + \text{production by reaction} = 0$$

If $F_{a0} = v_0 C_{a0}$ is the molar feed rate of component A into the reactor shown in Figure K1, we have

$$\text{input of A} = F_{a0}(1 - X_{a0}) = F_{a0} \text{ mol/time}$$

$$\text{output of A} = F_a = F_{a0}(1 - X_a) \text{ mol/time}$$

$$\text{production of A} = r_a V \text{ mol/time}$$

or

$$F_{a0} - F_{a0}(1 - X_a) + r_a V = 0$$

which reduces to

$$F_{a0} X_a = (-r_a) V$$

Rearranging,

$$\frac{V}{F_{a0}} = \frac{V}{v_0 C_{a0}} = \frac{X_a}{(-r_a)}$$

or

$$\tau = \frac{1}{s} = \frac{V}{v_0} = \frac{C_{a0} X_a}{(-r_a)}$$

Note that X_a and $(-r_a)$ are evaluated at exit stream conditions, which are the same as the conditions within the reactor (by definition). If the feed enters the reactor partially converted, we could rewrite the equation as

$$\frac{V}{F_{a0}} = \frac{V}{v_0 C_{a0}} = \frac{X_{af} - X_{ai}}{(-r_a)_f}$$

$$\tau = \frac{1}{s} = \frac{V}{v_0} = \frac{V C_{a0}}{F_{a0}} = \frac{C_{a0}(X_{af} - X_{ai})}{(-r_a)_f}$$

Thus the design equation for this reactor shows that knowing any three of the four terms X_a, $(-r_a)$, V, or F_{a0} gives the fourth term directly. So in kinetic experiments, each run at a given τ will give a corresponding value for the rate of

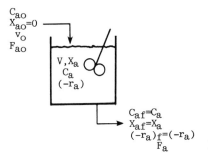

C_{a0}
$X_{a0} = 0$
v_0
F_{a0}

V, X_a
C_a
$(-r_a)$

$C_{af} = C_a$
$X_{af} = X_a$
$(-r_a)_f = (-r_a)$
F_a

FIGURE K1. Stirred tank reactor.

reaction directly. For this reason, the backmix reactor is frequently used for kinetic studies.

1. Space-Time and Space-Velocity

A good way of expressing the relation between the feed rate F_{a0} and reactor volume V in a flow system is to apply the terms space-time and space-velocity:

$$\text{space-time} = \tau = \frac{1}{s}$$

$$= \frac{\text{time required to process one reactor volume of feed}}{\text{measured at specified conditions}} = \text{time}$$

$$\text{space-velocity} = s = \frac{1}{\tau}$$

$$= \frac{\text{vol of entering feed at specified conditions/time}}{\text{void vol of reactor}} = (\text{time})^{-1}$$

We usually select the conditions as those of the entering feed, thus

$$\tau = \frac{1}{s} = \frac{C_{a0} V}{F_{a0}} = \left[\left(\frac{\text{mol A entering}}{\text{vol of feed}} \right) (\text{vol of reactor}) \right] \bigg/ \frac{(\text{mol A entering})}{(\text{time})}$$

$$= \frac{V}{v_0} = \frac{\text{reactor vol}}{\text{volumetric feed rate}}$$

C. Steady-State Plug Flow Reactor

In a plug flow reactor the composition of the fluid varies from point to point along the flow path; thus the material balance for a component must be made for a differential element of volume dV as shown in Figure K2. So for component A we have

$$\text{input-output} + \text{production by reaction} = 0$$

For a differential volume of dV,

$$\text{input of A} = F_a \text{ mol/time}$$

$$\text{output of A} = F_a + dF_a \text{ mol/time}$$

$$\text{production of A} = r_a \, dV \text{ mol/time}$$

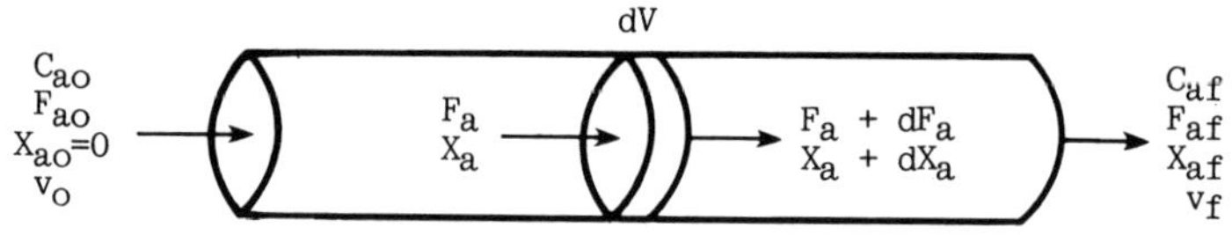

FIGURE K2. Plug flow reactor.

So

$$F_a - F_a - dF_a + r_a \, dV = 0$$

but

$$F_a = F_{a0}(1 - X_a)$$

then

$$dF_a = -F_{a0} \, dX_a$$

so

$$F_{a0} \, dX_a = -r_a \, dV$$

This is the balance over a differential slice of the total volume. For the whole reactor, we must integrate

$$\frac{1}{F_{a0}} \int_0^V dV = \frac{V}{F_{a0}} = \int_0^{X_{af}} \frac{dX_a}{(-r_a)}$$

or

$$\tau = \frac{1}{s} = \frac{V}{v_0} = C_{a0} \int_0^{X_{af}} \frac{dX_a}{(-r_a)}$$

Notice that for the plug flow reactor $(-r_a)$ varies while for the backmix reactor it is constant. If the feed enters partially converted, the equation would be

$$\tau = \frac{1}{s} = \frac{V}{v_0} = C_{a0} \int_{X_{ai}}^{X_{af}} \frac{dX_a}{(-r_a)}$$

The design equations for all reactor types are summarized in table K2.

TABLE K2 Ideal Reactor Design Equations

Reactor Type	General Equation
Batch[a]	$t = N_{a0} \int \dfrac{dX_a}{V(-r_a)}$
Plug flow	$\tau = C_{a0} \int \dfrac{dX_a}{-r_a}$
Stirred tank	$\tau = \dfrac{C_{a0} X_a}{-r_a}$

[a] For changing volume, use

$$V = V_0(1 + \varepsilon_a X_a)$$

PROBLEMS

A. For the reaction stoichiometry

$$A + B \rightarrow R, \qquad R + B \rightarrow S$$

a batch reactor gave the following test data:

	(min)							
Time	0	10	20	30	40	50	60	70
C_a (g mol/L)	1.40	0.89	0.64	0.47	0.37	0.29	0.23	0.19
C_b (g mol/L)	3.00	2.41	2.07	1.80	1.61	1.45	1.32	1.22

If you can show that the rate equations

$$-r_a = -\frac{dC_a}{dt} = k_1 C_a C_b, \qquad r_s = \frac{dC_s}{dt} = k_2 C_b C_r$$

are consistent with these data, what are the values for k_1 and k_2?

B. A gas decomposes at 900°C according to

$$2A(g) \rightarrow 2R(g) + S(g)$$

with a rate constant of 1000 cm^3/g mol sec using pure A.

(a) How long will it take to react 80% of A in a batch reactor at 900°C and 1 atm?

(b) What plug flow reactor volume, at 900°C and 3 atm, will be required to react 80% of A with a feed rate of 500 lb/hr? Molecular weight of A = 45.

C. We wish to produce D by reacting A over a solid catalyst:

$$A(g) \rightleftarrows D(g) + E(g)$$

Running at 0.2 atm and 1000°F, how many pounds of catalyst are needed to get a 30% conversion of A using a feed of 10 lb mol of pure A per hour?

$$r_a = K_i\left(p_a - \frac{p_d p_e}{K}\right)\bigg/(1 + K_a p_a + K_{de} p_{de})^2 = \frac{\text{lb mol A converted}}{\text{(lb catalyst)(hr)}}$$

p = partial pressure atm; $p_{de} = (p_d + p_e)/2$;

$K_i = 0.070$; $K_a = 0.40$; $K_{de} = 1.50$; $K = 0.25$.

D. The isothermal, irreversible, aqueous phase reaction

$$A + B = E$$

at 100°F obeys

$$\frac{dC_e}{dt} = r_e = kC_aC_b, \qquad k = 15 \frac{ft^3}{lb\ mol\ hr}$$

Using a 1000-ft^3 stirred tank reactor with an aqueous feed of 2000 ft^3/hr, what will be the outlet concentration of E if the inlet concentration of A and B are both 0.25 lb mol/ft^3?

E. A constant-volume batch reactor is used to get the following data on the decomposition of A:

	C_a (g mol/L)							
	15.0	12.0	10.5	7.5	6.0	4.5	3.0	1.5
At 100°C, time of run (sec)	0	56	89	173	229	300	402	573
At 120°C, time of run (sec)	0	28	—	86	—	150	201	287
At 150°C, time of run (sec)	0	11	—	35	—	61	82	116

Using this data, determine the reaction order and the Arrhenius constant.

F. The liquid-phase hydrolysis of dilute aqueous acetic anhydride solution

$$(CH_3CO)_2O + H_2O \longrightarrow 2CH_3COOH$$

is second-order irreversible (according to kinetic books). Our data on rates as a function of temperature were

$t(°C)$	10°	15°	25°	40°
Rate (g mol/cm^3 min)	0.0567C	0.0806C	0.158C	0.380C

where C = acetic anhydride concentration (g mol/cm^3).

Why do these data indicate that the reaction is first order? What is the Arrhenius constant?

PROBLEM-SOLVING STRATEGIES

A. My first reaction to this reaction problem would be to try to solve the two differential equations for C_a and C_b as a function of time. The k's could then be calculated at each experimental point. But because these are second-order reactions, the differential equations may not have easy, analytical solutions. Because of time limitations it is probably better to do numerical differentiation directly. To do this we first need to do a material balance on the stoichiometry equations to relate C_s and C_r to C_a and C_b, the experimental data. Then for each time period, we would use

$$\Delta C_a/10 = k_1(C_a C_b)_{av}$$

to calculate k_1. If the calculated k_1 values do not vary much, the rate expressions are consistent with the experimental data.

B. This gaseous reaction problem is variable volume due to the stoichiometry. So use the ε concept to relate volume to conversion. Inserting the reaction rate equation into the variable-volume batch reactor equation and performing the integration analytically should provide the solution for part (a). To get the initial concentration, use $PV = nRT$ or $C = P/RT$. For part (b), the rate equation is inserted into the variable-volume plug flow reactor equation and integrated for the solution. Because we do not know upon which component the rate constant is based, we must specify one for the sake of the test grader.

C. This catalytic reaction problem is just a plug flow reactor; in place of reactor volume we have pounds of catalyst. We must relate the partial pressures to conversion using the reaction stoichiometry. If the integration can be done analytically, it would be better than a numerical or graphical integration.

D. Aqueous phase reaction means that we can assume that the reaction is constant volume. The solution results from inserting the rate expression into the stirred tank equation.

E. Assume that the rate expression is

$$r_a = -kC_a^n$$

where n is unknown. Then, since for a constant volume batch reactor,

$$r_a = \frac{dC_a}{dt} = -kC_a^n$$

we integrate

$$\int_{C_{a1}}^{C_{a2}} \frac{dC_a}{C_a^n} = -k \int_{t_1}^{t_2} dt$$

Bear in mind that $n = 1$ must be integrated differently from other n's. The reason for integrating from t_1 to t_2 is that we notice that some of the concentrations are just doubled. So we could choose several different sets of C_{a1} and C_{a2} at each temperature. If these sets give the same k, we have assumed the correct n. Once we have the correct n, we know the k at each temperature. then from

$$\ln k = \ln k_0 - (E/RT)$$

we can calculate k_0 and E/R.

F. For this second-order reaction,

$$r_p = kC_a C_{H_2O}$$

If the acetic acid is diluted, the concentration of water is large and small changes in water concentration will have negligible effect on C_{H_2O}—it is constant. This problem is a gift.

SOLUTIONS

A. $A + B \rightarrow R$ $R + B \rightarrow S$

C_{a0} C_{b0} 0 C_{r1} C_{b1} 0

C_a C_{b1} C_{r1} C_r C_b C_s

$C_{a0} - C_a = C_{b0} - C_{b1} = C_{r1}$ $C_{r1} - C_r = C_{b1} - C_b = C_s$

$C_{r1} = C_{a0} - C_a,$ $C_{a0} - C_a - C_r$

$\quad C_{b1} = C_{b0} - C_{a0} + C_a;$ $\quad = C_{b0} - C_{a0} + C_a - C_b = C_s$

So

$$C_s = (C_{b0} - C_b) - (C_{a0} - C_a), \qquad C_r = 2(C_{a0} - C_a) - (C_{b0} - C_b)$$

t (min)	0	10	20	30	40	50	60	70
C_b	3.00	2.41	2.07	1.80	1.61	1.45	1.32	1.22
C_a	1.40	0.89	0.64	0.47	0.37	0.29	0.23	0.19
$C_{b0} - C_b$	0	0.59	0.93	1.20	1.39	1.55	1.68	1.78
$C_{a0} - C_a$	0	0.51	0.76	0.93	1.03	1.11	1.17	1.21
C_s	0	0.08	0.17	0.27	0.36	0.44	0.51	0.57
C_r	0	0.43	0.59	0.66	0.67	0.67	0.66	0.64
$C_a C_b$	4.20	2.14	1.32	0.85	0.60	0.42	0.30	0.23
$C_b C_r$	0	1.04	1.22	1.19	1.08	0.97	0.87	0.78

t average (min)	5	15	25	35	45	55	65	
$(C_a C_b)_{av}$	3.17	1.73	1.09	0.73	0.51	0.36	0.27	
$(C_b C_r)_{av}$	0.52	1.13	1.21	1.14	1.03	0.92	0.83	
$-\Delta C_a/\Delta t$	0.0510	0.0250	0.0170	0.0100	0.0080	0.0060	0.0040	
$\Delta C_s/\Delta t$	0.0080	0.0090	0.0100	0.0090	0.0080	0.0070	0.0060	
k_1	0.0161	0.0145	0.0156	0.0137	0.0157	0.0167	0.0148	$(0.0153)_{av}$
k_2	0.0154^a	0.0080	0.0083	0.0079	0.0078	0.0076	0.0072	$(0.0078)_{av}$

$$k_1 = -\left(\frac{\Delta C_a}{\Delta t}\right)\bigg/ C_a C_b, \qquad k_2 = \left(\frac{\Delta C_s}{\Delta t}\right)\bigg/ C_b C_r$$

$$k_1 = 0.0153 \text{ L/g mol min}, \qquad k_2 = 0.0078 \text{ L/g mol min}$$

[a] Ignored since it is most likely in error.

B. (a) $2A \rightarrow 2R + S$, $k = 10^3$ cm^3/g mol sec $r_a = -kC_a^2$
Batch reactor with $V = V_0(1 + \varepsilon x)$.'

$$\varepsilon = \frac{V_{x=1} - V_{x=0}}{V_{x=0}} = \frac{3 - 2}{2} = \tfrac{1}{2}; \qquad C_a = \frac{C_{a0}(1 - x)}{1 + \varepsilon x};$$

$$\frac{C_{a0}}{1 + \varepsilon x}\frac{dx}{dt} = +\frac{kC_{a0}^2(1 - x)^2}{(1 + \varepsilon x)^2}$$

Assume k based upon A, as written

$$\int_0^t k\, dt = \int_0^x \frac{C_{a0}(1 + \varepsilon x)^2\, dx}{(1 + \varepsilon x)C_{a0}^2(1 - x)^2} = \int_0^x \frac{1 + \varepsilon x}{C_{a0}(1 - x)^2}\, dx$$

$$C_{a0} kt = \int_0^x \frac{dx}{(1 - x)^2} + \varepsilon \int_0^x \frac{x\, dx}{(1 - x)^2}$$

$$= \left[\frac{1}{1 - x}\right]_0^x + \varepsilon\left[\ln(1 - x) + \frac{1}{1 - x}\right]_0^x$$

$$= \frac{1}{1 - x} - 1 + \varepsilon\left[\ln(1 - x) + \frac{1}{1 - x} - 1\right]$$

$$= \frac{1 - 1 + x}{1 - x}(1 + \varepsilon) + \varepsilon \ln(1 - x)$$

$$= \frac{(1 + \varepsilon)x}{1 - x} + \varepsilon \ln(1 - x)$$

$$PV = nRT \quad \text{or} \quad \frac{n}{V} = C = \frac{P}{RT}; \quad C_{a0} = \frac{1 \text{ atm}}{82.06(1173°K)}$$

$$t = \frac{82.06(1173)}{1000} \left[\frac{\frac{3}{2}(0.8)}{0.2} + \frac{1}{2} \ln (0.2) \right] = 500.08 \text{ sec} = 8.33 \text{ min}$$

(b) Plug flow

$$\tau = \frac{V}{v_0} = C_{a0} \int_0^x \frac{dx}{(-r)} = C_{a0} \int_0^x \frac{dx(1 + \varepsilon x)^2}{kC_{a0}^2(1 - x)^2}$$

$$\frac{V}{v_0} = \frac{1}{kC_{a0}} \int_0^x \frac{(1 + \varepsilon x)^2}{(1 - x)^2} dx$$

$$= \frac{1}{kC_{a0}} \left[\int_0^x \frac{dx}{(1 - x)^2} + 2\varepsilon \int_0^x \frac{x \, dx}{(1 - x)^2} + \varepsilon^2 \int_0^x \frac{x^2 \, dx}{(1 - x)^2} \right]$$

$$= \frac{1}{kC_{a0}} \left\{ \frac{1}{1 - x} + 2\varepsilon \left[\ln (1 - x) + \frac{1}{1 - x} \right] \right.$$

$$\left. - \varepsilon^2 \left[1 - x - 2 \ln (1 - x) - \frac{1}{1 - x} \right] \right\}_0^{0.8}$$

$$= \frac{1}{kC_{a0}} \left[\frac{(1 + 2\varepsilon + \varepsilon^2)}{1 - x} + 2\varepsilon(1 + \varepsilon) \ln (1 - x) - \varepsilon^2(1 - x) \right]_0^{0.8}$$

$$C_{a0} = \frac{3 \text{ atm}}{82.06(1173)}$$

$$\frac{V}{v_0} = \frac{82.06(1173)}{3(1000)} \left[\frac{1 + 2(\frac{1}{2}) + \frac{1}{4}}{0.2} \right.$$

$$\left. + 2(\tfrac{1}{2})(\tfrac{3}{2}) \ln 0.2 - \tfrac{1}{4}(0.2) - \frac{(1 + 1 + \frac{1}{4})}{1} + \tfrac{1}{4}(1) \right]$$

$$\frac{V}{v_0} = \frac{82.06(1173)}{3(1000)} \left(\frac{\frac{9}{4}}{0.2} + \tfrac{3}{2} \ln 0.2 - \frac{0.2}{4} - \tfrac{9}{4} + \tfrac{1}{4} \right)$$

$$= 217.73 \text{ sec} = 3.63 \text{ min} = 0.0605 \text{ hr}$$

$$v_0 = \frac{\text{ft}^3 \text{ feed}}{\text{hr}} = \frac{500 \text{ lb}}{\text{hr}} \frac{1 \text{ lb mol}}{45 \text{ lb}} \frac{454 \text{ g}}{\text{lb}} \frac{82.06(1173)\text{cm}^3}{3 \text{ g mol}} \frac{1}{10^3 \text{ cm}^3} \frac{\text{ft}^3}{28.3 \text{ L}}$$

$$= 5719.2 \text{ ft}^3/\text{hr}$$

$$V = v_0(0.0605 \text{ hr}) = 5719.2 \text{ ft}^3/\text{hr} (0.0605) = 346 \text{ ft}^3$$

C.

$$
\begin{array}{ccccc}
 & & A & \rightleftarrows & D & + & E \\
t = 0 & & A_0 & & 0 & & 0 \\
t = t & & A_0(1 - x) & & A_0 x & & A_0 x
\end{array}
$$

$$
\text{mf} \qquad \frac{1 - x}{1 + x} \qquad \frac{x}{1 + x} \qquad \frac{x}{1 + x}
$$

$$
p \qquad \frac{1 - x}{1 + x} P \qquad \frac{xP}{1 + x} \qquad \frac{xP}{1 + x}
$$

$\sum = A_0(1 + x)$; $P = $ total pressure; atm $= 0.2$

$$
p_{de} = \frac{xP}{1 + x}
$$

$$
\int_0^x -\frac{dx}{r_a} = \int_0^w \frac{dw}{F_{a0}}, \qquad r_a = \frac{\text{lb mol A reacted}}{\text{(lb catalyst) (hr)}}
$$

So

$$
-r_a = r_a \text{ as given}
$$

$$
F_{a0} = \frac{\text{lb mol A feed}}{\text{hr}}
$$

$$
-r_a = K_i \left(\frac{1 - x}{1 + x} P - \frac{x^2 P^2}{K(1 + x)^2} \right) \Big/ \left(1 + K_a P \frac{1 - x}{1 + x} + K_{de} \frac{xP}{1 + x} \right)^2
$$

$$
= K_i \left[(1 - x)(1 + x) - \frac{x^2 P}{K} \right] P \Big/ [1 + x + K_a P(1 - x) + K_{de} xP]^2
$$

$$
-\frac{1}{r_a} = \frac{[(1 + K_a P) + (1 - K_a P + K_{de} P)x]^2}{K_i P[1 - (1 + P/K)x^2]} = \frac{(1.08 + 1.22x)^2}{0.014(1 - 1.8x^2)}
$$

or

$$
0.014 \frac{W}{F_{a0}} = \int_0^x \frac{(1.08 + 1.22x)^2}{1 - 1.8x^2} dx = \int_0^x \frac{1.1664 + 2.6352x + 1.4884x^2}{1 - 1.8x^2} dx
$$

$$
= 1.1664 \left[\frac{1}{2\sqrt{1.8}} \ln \frac{1 + \sqrt{1.8}\,x}{1 - \sqrt{1.8}\,x} \right]_0^x
$$

$$
+ 2.6352 \left[-\frac{1}{20.8} \ln (1 - 1.8x^2) \right]_0^x
$$

$$
+ 1.4884 \left[-\frac{x}{1.8} + \frac{1}{2(1.8)\sqrt{1.8}} \ln \frac{1 + \sqrt{1.8}\,x}{1 - \sqrt{1.8}\,x} \right]_0^x
$$

$$= \frac{1.1664}{2\sqrt{1.8}} \ln \frac{1 + \sqrt{1.8}x}{1 - \sqrt{1.8}x} - \frac{2.6352}{2(1.8)} \ln(1 - 1.8x^2) + \frac{1.4884}{1.8}$$

$$\times \left(-x + \frac{1}{2\sqrt{1.8}} \ln \frac{1 + \sqrt{1.8}x}{1 - \sqrt{1.8}x} \right)$$

$$0.014 \frac{W}{F_{a0}} = 0.5151$$

So

$$W = \frac{0.5151}{0.014} F_{a0} = \frac{0.5151}{0.014} 10 = 368 \text{ lb}$$

D. $A + B \xrightarrow{15} E$, $\tau = \dfrac{C_{a0} X_a}{-r_a}$, $r_a = \left(\dfrac{\text{mol A reacted}}{\text{ft}^3 \text{ min}} \right)$ at exit conditions

$$\tau = \frac{V}{v_0} = \frac{1000 \text{ ft}^3}{2000 \text{ ft}^3/\text{hr}} = 0.5 \text{ hr}$$

$$-r_a = (kC_a C_b)_{\text{at exit}}, \qquad C_a = C_{a0}(1 - x_a) = C_b$$

$$-r_a = kC_{a0}^2(1 - x_a)^2 \qquad \text{so} \qquad 0.5 \text{ hr} = \frac{C_{a0} x_a}{kC_{a0}^2(1 - x_a)^2}$$

or

$$(1 - x_a)^2 = \frac{x_a}{0.5 \cdot 15 \cdot 0.25} = 0.5333 x_a$$

$$1 - 2x_a + x_a^2 = 0.5333 x_a \qquad \text{or} \qquad x_a^2 - 2.5333 x_a + 1 = 0$$

$$x_a = \frac{2.5333 \pm \sqrt{2.5333^2 - 4}}{2} = 2.0441 \qquad \text{or} \qquad 0.4892$$

Now

$$C_E = C_{a0} x_a = 0.1223 \text{ lb mol/ft}^3$$

E. For an nth-order reaction, $C_a^{1-n} - C_{a0}^{1-n} = k(n - 1)t$, $n \neq 1$; for a first-order reaction, $\ln(C_a/C_{a0}) = -kt$.

C_a/C_{a0}	1	0.8	0.5	0.3	0.2	0.1	T
t	0	56	173	300	402	573	at 100°C
$[-\ln(C_a/C_{a0})]/t$	—	0.00398	0.00401	0.00401	0.00400	0.00402	~0.00401
t	0	28	86	150	201	287	at 120°C
$[-\ln(C_a/C_{a0})]/t$	—	0.00797	0.00806	0.00803	0.00801	0.00802	~0.00802
t	0	11	35	61	82	116	at 150°C
$[-\ln(C_a/C_{a0})]/t$	0	0.02029	0.01980	0.01974	0.01963	0.01985	~0.01986

$$k = k_0 e^{-E/RT}, \qquad \ln k = \ln k_0 - (E/RT)$$

or

$$k_{150} = k_0 e^{-E'/423}, \qquad k_{120} = k_0 e^{-E'/393}, \qquad k_{100} = k_0 e^{-E'/373}$$

$$\frac{k_{150}}{k_{100}} = e^{-E'[(1/423)-(1/373)]} \qquad \text{or} \qquad E' = 5049$$

$$\hspace{10cm} 5065$$

$$\frac{k_{120}}{k_{100}} = e^{-E'[(1/393)-(1/373)]} \qquad \text{or} \qquad E' = 5080$$

$$k = k_0 e^{-5065/T}; \; k_0 = k e^{5065/T};$$

$k_0 = 3166, 3172, 3149,$ or 3162; so

$$r_a = -3162 e^{-5065/T} C_a$$

F. The stoichiometry indicates that the rate should probably be $R = k C_{H_2O} C_A$. When the reaction is run with large amounts of water C_{H_2O} is approximately constant and the rate looks like $R = \bar{k} C_a$, where $\bar{k} = k C_{H_2O}$,

$$\frac{R}{C_a} = \bar{k} = k_0 e^{-E/RT}, \qquad \ln \frac{R}{C_a} = \ln k_0 - \frac{E}{R} \frac{1}{T}$$

R/C_a	$1/T$	$\ln (R/C_a)$
0.0567	0.003534	−2.8700
0.0806	0.003472	−2.5183
0.158	0.003356	−1.8452
0.380	0.003195	−0.9676

$$-2.8700 = \ln k_0 - 0.003534(E/R)$$

$$-0.9676 = \ln k_0 - 0.003195(E/R)$$

$$E/R = 5612, \qquad \ln k_0 = 16.96210, \qquad k_0 = 2.326 \times 10^7$$

$$R = 2.326 \times 10^7 e^{-5612/T} \text{K}$$

FLUID FLOW

W. L. McCabe and J. C. Smith, *Unit Operations of Chemical Engineering*, McGraw-Hill, New York, 1976.

I. BASIC PRINCIPLES

A. Hydrostatic Equilibrium

In any cross section parallel to the surface of the earth, pressure is constant, varying only from height to height. The equation for a vertical column of fluid is

$$dp + \frac{g}{g_c} \rho \, dZ = 0$$

where Z = height above base of column,
p = pressure at this point,
ρ = density at this point,
g = acceleration of free fall in earth's gravitational field,
$\cong 32.274 \text{ ft/sec}^2$, and
g_c = Newton's law proportionality factor for the gravitational force unit, $\cong 32.174 \text{ lb ft/sec}^2$ lb force.

B. Total Energy Equation of Steady Flow

Consider Figure F1 for a single stream of material. An energy balance gives

$$m\left[\frac{u_b^2 - u_a^2}{2g_c J} + \frac{g(Z_b - Z_a)}{g_c J} + H_b - H_a\right] = Q - \frac{W_s}{J}$$

where J is the mechanical equivalent of heat = 778.17 ft lb$_f$/Btu

Some limitations of this equation are

- surface, magnetic, and so forth, energy changes are neglected;
- heat effects Q are + when heat flows *into* the equipment; and
- shaft work W_s is + when work is done *by* the equipment.

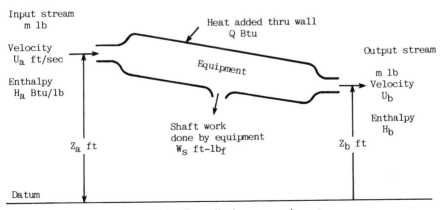

FIGURE F1. Generalized process equipment.

C. Manometers

Consider the manometer of Figure F2. Using the hydrostatic equilibrium equation between points 1 and 2 gives

$$\int_{p_2}^{p_a} dp = -\frac{g}{g_c} \rho_b \int_{Z_2}^{Z_a} dZ$$

between 5 and 4

$$\int_{p_4}^{p_b} dp = -\frac{g}{g_c} \rho_b \int_{Z_4}^{Z_b} dZ$$

between 4 and 3

$$\int_{p_3}^{p_4} dp = -\frac{g}{g_c} \rho_a \int_{Z_3}^{Z_4} dZ$$

Thus

$$p_a - p_2 = -\frac{g}{g_c} \rho_b (Z_a - Z_2); \qquad p_b - p_4 = -\frac{g}{g_c} \rho_b (Z_b - Z_4)$$

$$p_4 - p_3 = -\frac{g}{g_c} \rho_a (Z_4 - Z_3)$$

Now

$$Z_a - Z_2 = Z_m + R_m, \qquad Z_b - Z_4 = Z_m, \qquad Z_4 - Z_3 = R_m$$

Adding the p_b–p_4 equation to the p_4–p_3 equation gives

$$p_b - p_3 = -\frac{g}{g_c} (\rho_a R_m + \rho_b Z_m)$$

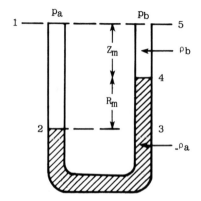

FIGURE F2. Manometer.

Now $p_3 = p_2$. So subtract this p_b-p_3 equation from the p_a-p_2 (which is now p_a-p_3) equation to obtain

$$p_a - p_b = \frac{g}{g_c} R_m(\rho_a - \rho_b)$$

the equation for a manometer, if p_a and p_b are measured in the same horizontal plane.

D.   Mass Balance

Consider the tube shown in Figure F3. At steady state, the mass rate of flow into the tube must equal the rate of mass out, considered over the period of time. Thus, if ρ and u are constant across area S,

$$\dot{m} = \rho_a u_a S_a = \rho_b u_b S_b$$

where $\dot{m}$ = rate of flow of mass (mass/time).

This is the equation of continuity, since

$$\dot{m} = \rho u S = \text{constant}$$

E.   Average Velocity

If the velocity u is not constant across the area S, we need an average velocity. Now if we look at a small part of the cross-sectional area S, we see that

$$d\dot{m} = \rho u \, dS$$

and then the total mass flow rate through S would be

$$\dot{m} = \rho \int_S u \, dS$$

assuming that density is constant across S. Then the average velocity $\bar{V}$ would be defined by

$$\bar{V} \equiv \frac{\dot{m}}{\rho S} = \frac{1}{S} \int_S u \, dS$$

Of course, we could also write the average velocity in terms of the total volumetric flow rate $\dot{V}$,

$$\bar{V} = \frac{\dot{V}}{S}$$

Velocity u_a
Density ρ_a
Area S_a

Velocity u_b
Density ρ_b
Area S_b

FIGURE F3.   Generalized flow tube.

F.   Mass Velocity

We can rewrite the average velocity in terms of a mass velocity G,

$$\bar{V}\rho = \frac{\dot{m}}{S} = G$$

G.   Macroscopic Momentum Balance

If we assume steady flow in the x direction, the sum of all forces acting upon the fluid in the x direction must equal the increase in the time rate of momentum of the flowing fluid, or

$$\sum F = \frac{1}{g_c}(\dot{M}_b - \dot{M}_a)$$

where $\dot{M}$ = momentum flow rate.

H.   Momentum of Total Stream

The momentum flow rate $\dot{M}$ of a fluid with a mass flow rate $\dot{m}$ and a velocity u is

$$\dot{M} = \dot{m}u$$

if u is constant. If u is not constant across S, we must use

$$\frac{d\dot{M}}{dS} = \rho u \cdot u = \rho u^2$$

or, for a constant density stream,

$$\frac{\dot{M}}{S} = \frac{\rho}{S}\int_S u^2\, dS$$

Using a momentum correction factor β,

$$\beta \equiv \frac{(\dot{M}/S)}{(\rho \bar{V}^2)}$$

or

$$\beta = \frac{1}{S\bar{V}^2}\int_S u^2\, dS = \frac{1}{S}\int_S \left(\frac{u}{\bar{V}}\right)^2 dS$$

If we know β, then

$$\dot{M} = \beta \bar{V}\dot{m}$$

and our momentum balance becomes

$$\sum F = \frac{\dot{m}}{g_c}(\beta_b \bar{V}_b - \beta_a \bar{V}_a)$$

I. Bernoulli Equation without Friction

Using a momentum balance for a steady potential flow, we obtain the point form of the Bernoulli equation without friction

$$\frac{1}{\rho}\frac{dp}{dL} + \frac{g}{g_c}\frac{dZ}{dL} + \frac{d(u^2/2)}{g_c\,dL} = 0$$

The differential form is

$$\frac{dp}{\rho} + \frac{g}{g_c}dZ + \frac{1}{g_c}d\frac{u^2}{2} = 0$$

Of course, between two points a and b, this can be integrated to

$$\frac{p_a}{\rho} + \frac{gZ_a}{g_c} + \frac{u_a^2}{2g_c} = \frac{p_b}{\rho} + \frac{gZ_b}{g_c} + \frac{u_b^2}{2g_c}$$

J. Mechanical Energy Equation

The Bernoulli equation (Section I) is just an energy balance; all terms have the dimensions of energy per unit mass. Mechanical potential is gZ/g_c; mechanical kinetic energy is $u^2/2g_c$; p/ρ is the mechanical work done to get the fluid into a tube or the work recovered as it leaves the tube.

K. The Kinetic Energy Term

The Bernoulli equation was developed for a system with u constant across a cross section. Instead, we could use a kinetic energy correction factor α in the kinetic energy term;

$$\alpha\frac{\bar{V}^2}{2g_c}$$

where

$$\alpha = \frac{\int_S u^3\,dS}{\bar{V}^3 S}$$

and $\bar{V}$ is the average velocity.

L. Fluid Friction Correction

For incompressible fluids, the friction form of the Bernoulli equation is

$$\frac{p_a}{\rho} + \frac{gZ_a}{g_c} + \frac{\alpha_a\bar{V}_a^2}{2g_c} = \frac{p_b}{\rho} + \frac{gZ_b}{g_c} + \frac{\alpha_b\bar{V}_b^2}{2g_c} + h_f$$

where h_f is the friction generated in the fluid between a and b per unit mass of fluid.

M. Bernoulli and Pump Work

We consider a pump between points a and b. The work supplied to the pump is shaft work W_s. Pump work is negative, due to shaft work sign convention. So pump work W_p is defined by

$$W_p = -\frac{W_s}{m}$$

Friction occurring within the pump must be accounted for. The mechanical energy supplied to the pump as negative shaft work must be discounted by the frictional losses to get the net mechanical energy available to the flowing fluid. If h_{fp} is the total friction in the pump per unit mass of fluid, then net work to the fluid is $W_p - h_{fp}$. But usually a pump efficiency η is used. This efficiency is defined by

$$\eta = (W_p - h_{fp})/W_p$$

Then the final Bernoulli equation for incompressible fluid flow is

$$\frac{p_a}{\rho} + \frac{gZ_a}{g_c} + \frac{\alpha_a \bar{V}_a^2}{2g_c} + \eta W_p = \frac{p_b}{\rho} + \frac{gZ_b}{g_c} + \frac{\alpha_b \bar{V}_b^2}{2g_c} + h_f$$

II. INCOMPRESSIBLE FLUID FLOW IN PIPES

A. Skin Friction and Wall Shear

If h_{fs} denotes the skin friction between the fluid stream and the wall, it can be shown that

$$h_{fs} = \frac{2\,\tau_w}{\rho\,r_w}\,\Delta L = \frac{4\,\tau_w}{\rho\,D}\,\Delta L$$

where τ_w = the shear stress at the wall (lb_f/ft^2),
   h_{fs} = skin friction (ft lb_f/lb), and
   D = pipe diameter (ft).

B. Friction Factor

The friction factor is defined to be the ratio of the shear stress at the wall to the product of density and velocity head, that is,

$$f \equiv \frac{\tau_w}{\rho \bar{V}^2/2g_c} = \frac{2g_c \tau_w}{\rho \bar{V}^2}$$

Measurement of skin friction in pipes can be done with h_{fs}, Δp_s, τ_w, or f. They are related by

$$h_{fs} = \frac{2\,\tau_w}{\rho\,r_w}\,\Delta L = -\frac{\Delta p_s}{\rho} = 4f\,\frac{\Delta L}{D}\,\frac{\bar{V}^2}{2g_c}$$

so

$$f = \frac{\Delta p_s}{2\Delta L} \frac{g_c D}{\rho \bar{V}^2}$$

Note that the friction factor f used here is the Fanning friction factor.

C. Laminar Flow in Pipes

1. Newtonian Fluids

Velocity as a Function of Radius. Using the definition of viscosity μ,

$$\mu = -\frac{\tau g_c}{du/dr}$$

we can find the equation for velocity

$$u = \frac{\tau_w g_c}{2r_w \mu} (r_w^2 - r^2)$$

from which we can find

$$u_{max} = \frac{\tau_w g_c r_w}{2\mu} \quad \text{and} \quad \frac{u}{u_{max}} = 1 - \left(\frac{r}{r_w}\right)^2$$

Average Velocity

$$\bar{V} = \frac{\tau_w g_c r_w}{4\mu} \quad \text{and} \quad \frac{\bar{V}}{u_{max}} = \frac{1}{2}$$

Kinetic Energy Correction

$$\alpha = 2$$

Momentum Correction

$$\beta = \frac{4}{3}$$

2. Non-Newtonian Fluids

The velocity profile shape for non-Newtonians differs from that of Newtonians. Many must be determined experimentally. A few simpler ones have analytical expressions. Advanced fluid flow texts should be consulted for non-Newtonian systems.

D. Turbulent Flow in Pipes

Turbulent flow is not nearly so easy to analyze as laminar. Much experimental and theoretical work has been expended. Some of the more useful ideas are discussed as follows.

1. Velocity Distribution

Turbulent flow velocity distribution has been expressed in terms of dimensionless parameters:

$$u^* = \text{friction velocity} \equiv \bar{V}\sqrt{\frac{f}{2}} = \sqrt{\frac{\tau_w g_c}{\rho}}$$

$$u^+ = \text{velocity quotient} \equiv \frac{u}{u^*}$$

$$y^+ = \text{distance} \equiv \frac{y u^* \rho}{\mu} = \frac{y}{\mu}\sqrt{\tau_w g_c \rho}$$

$$y = \text{distance from tube wall with } r_w = r + y$$

2. Universal Velocity-Distribution Equations

Laminar Sublayer—Near the Wall

$$u^+ = y^+$$

Buffer Layer—between Sublayer and Turbulent Core

$$u^+ = 5 \ln y^+ - 3.05 \text{ (empirical equation)}$$

Turbulent Core

$$u^+ = \frac{1}{k}\ln y^+ + c_1$$

with $k = 0.407$, $c_1 = 5.67$.

Sublayer

$$y^+ < 5; \qquad \text{buffer } 5 < y^+ < 30; \qquad \text{turbulent core } 30 < y^+$$

3. Flow Quantities

Since the sublayer and the buffer layer are only a small part of the flow regime, the turbulent core equation may be used, to a first approximation, to determine the various flow quantities:

Average Velocity

$$\frac{\bar{V}}{u_{max}} = \frac{1}{1 + (3/2k)\sqrt{f/2}}$$

Friction Factor-Reynolds Number Relationship

$$\frac{1}{\sqrt{f/2}} = \frac{1}{k}\ln\left(N_{Re}\sqrt{\frac{f}{8}}\right) - \frac{3}{2k} + c_1$$

$$N_{Re} = \frac{D\bar{V}}{v} = \frac{D\bar{V}\rho}{\mu}$$

Corrections for Kinetic Energy and Momentum

$$\alpha = 1 + \frac{f}{8k^2}\left(15 - \frac{9}{k}\sqrt{f}\right)$$

$$\beta = 1 + \frac{5}{8k^2}f$$

but α, β are both almost one.

The Constants k and c_1

$k = 0.407$, $c_1 = 5.67 \equiv$ Nikuradse equation, that is,

$$\frac{1}{\sqrt{f}} = 4.0 \log\left(N_{\text{Re}}\sqrt{f}\right) - 0.60$$

4. Friction Factors, Noncircular Cross Section

Long straight channels with constant noncircular cross section can be handled with the previous friction equations, if we use the equivalent diameter as the diameter in the Reynolds number and in the friction factor definition. The equivalent diameter for a circular tube is defined to be four times the hydraulic radius r_{H}.

$$r_{\text{H}} = \frac{S}{LP}$$

where $S =$ cross-sectional area of channel and
$LP =$ perimeter of wetted channel.

Note that hydraulic radius should not be used with laminar flow.

5. Friction with Direction or Velocity Changes

Friction Loss from Sudden Expansion. (See Figure F4 for nomenclature.)

$$h_{fe} = K_e\left(\frac{\bar{V}_a^2}{2g_c}\right)$$

$$= \text{friction loss}$$

$$K_e = \left(1 - \frac{S_a}{S_b}\right)^2$$

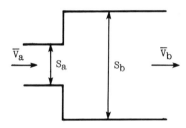

FIGURE F4. Sudden expansion.

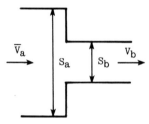

FIGURE F5. Sudden contraction.

Friction Loss from Sudden Contraction. (See Figure F5 for nomenclature.)

$$h_{fc} = K_c\left(\frac{\bar{V}_b^2}{2g_c}\right)$$

$$K_c = 0.4\left(1 - \frac{S_b}{S_a}\right) \qquad \text{(empirical)}$$

Effect of Valves and Fittings

$$h_{fc} = K_f\left(\frac{\bar{V}_a^2}{2g_c}\right), \qquad \bar{V}_a = \text{average velocity prior to fitting}$$

where

	K_f
Globe valve, wide open	10.0
Angle valve, wide open	5.0
Gate valve, wide open	0.2
half-open	5.6
Return bend	2.2
Tee	1.8
Elbow, 90°	0.9
45°	0.4

Bernoulli Equation and Form-Friction Losses
To include all form-friction terms in Bernoulli's equation, we use

$$h_f = \left[4f\left(\frac{L}{D}\right) + K_c + K_e + K_f\right]\frac{\bar{V}^2}{2g_c}$$

where the first term accounts for the skin friction loss.

III. FLOW OF COMPRESSIBLE FLUIDS

A. General Information

We now enter the realm of density variations. In compressible flow at regular densities and high velocities, a basic parameter is the Mach number, the ratio of the speed of the fluid to the speed of sound in the fluid under conditions of flow, $N_{\text{Ma}} = u/a$. The following simplifying assumptions are used: the flow is steady and unidirectional; velocity is constant across a cross-sectional area; shaft work is zero; the only friction considered is wall shear; gravitational effects and mechanical energy are neglected; and the fluid is considered an ideal gas with a constant specific heat.

1. Continuity Equation

Previously

$$\ln \rho + \ln S + \ln u = \text{constant}$$

and so

$$\frac{d\rho}{\rho} + \frac{dS}{S} + \frac{du}{u} = 0$$

2. Total Energy Balance

$$\frac{Q}{m} = H_b - H_a + \frac{u_b^2}{2g_c J} - \frac{u_a^2}{2g_c J}$$

$$\frac{dQ}{m} = dH + d\left(\frac{u^2}{2g_c J}\right)$$

3. Mechanical Energy Balance

$$\frac{dp}{\rho} + d\left(\frac{u^2}{2g_c}\right) + \frac{u^2}{2g_c} \cdot \frac{f \, dL}{r_H} = 0$$

4. Velocity of Sound

$$a = \sqrt{g_c \left(\frac{dp}{d\rho}\right)_S}$$

where the subscript S means at constant entropy.

5. Ideal-Gas Equations

$$p = \frac{R}{M}\rho T$$

$$\frac{dp}{p} = \frac{d\rho}{\rho} + \frac{dT}{T}$$

$$H = H_0 + C_p(T - T_0)$$

$$dH = C_p \, dT$$

$$\left.\begin{array}{l} p\rho^{-\gamma} = \text{constant} \\ Tp^{-(1-1/\gamma)} = \text{constant} \end{array}\right\} \text{isentropic}$$

$$\gamma = \frac{C_p}{C_v} = \frac{C_p}{C_p - (R/MJ)}$$

$$\frac{dp}{p} = \gamma \frac{d\rho}{\rho}, \qquad \left(\frac{dp}{d\rho}\right)_s = \gamma \frac{p}{\rho}$$

$$a = \sqrt{\frac{g_c\gamma p}{\rho}} = \sqrt{\frac{g_c\gamma TR}{M}}$$

$$N_{\text{Ma}}^2 = \frac{\rho u^2}{g_c\gamma p} = \frac{u^2 M}{g_c\gamma TR}$$

6. Stagnation Temperature

If a fluid is brought to rest adiabatically without shaft work, the temperature attained is the stagnation temperature T_s

$$T_s = T + \frac{u^2}{2g_c J C_p}$$

and the stagnation enthalpy H_s is

$$H_s = H + \frac{u^2}{2g_c J}$$

B. Compressible Flow Process

1. Introduction

Assume a large supply of gas at specified temperature and pressure at zero velocity. This gas is in a reservoir at reservoir conditions (stagnation temperature). We assume the gas flows from the reservoir, without entrance friction, into and through some conduit. Our gas leaves this conduit at definite temperature, pressure, and velocity into the exhaust reservoir. Here the pressure may be controlled at a constant value less than the reservoir pressure. Several processes may occur within the conduit.

a. Isentropic Expansion

Cross-sectional area of conduit must change. Stagnation temperature is constant in the conduit.

b. Adiabatic Frictional Flow

Cross-sectional area constant, stagnation temperature constant; irreversible.

c. Isothermal Frictional Flow

Cross-sectional area constant, heat flow through the wall to keep the temperature constant.

2. Variable Area Conduits

a. Introduction

An isentropic flow device is a nozzle, illustrated in Figure F6. It usually consists of a convergent section, a throat section, and a divergent section. These devices are frequently used in the measurement of fluid flow.

b. Isentropic Flow Equations

Gas Properties Change during Flow

$$\frac{p}{\rho^\gamma} = \frac{p_0}{\rho_0^\gamma}; \frac{T}{p^{1-(1/\gamma)}} = \frac{T_0}{p_0^{1-(1/\gamma)}}$$

Subscript 0 indicates reservoir conditions.

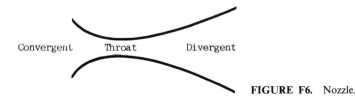

Convergent   Throat   Divergent

FIGURE F6. Nozzle.

Nozzle Velocity

$$u^2 = \frac{2\gamma g_c p_0}{(\gamma - 1)\rho_0}\left[1 - \left(\frac{p}{p_0}\right)^{1-(1/\gamma)}\right]$$

or

$$N_{\text{Ma}}^2 = \frac{2}{\gamma - 1}\left[\left(\frac{p_0}{p}\right)^{1-(1/\gamma)} - 1\right]$$

from which

$$\frac{p}{p_0} = 1 \bigg/ \left\{1 + \left[\frac{\gamma - 1}{2}\right]N_{\text{Ma}}^2\right\}^{\gamma/(\gamma - 1)}$$

When $u = a$ and $N_{\text{Ma}} = 1$, the critical pressure ratio r_c is

$$r_c = \frac{p(u = a, N = 1)}{p_0} = \left(\frac{2}{\gamma + 1}\right)^{\gamma/(\gamma - 1)}$$

c. Cross-Sectional Area Effects

The changes in velocity and Mach number caused by area changes can be determined from

$$\frac{du}{u}\left(1 - \frac{u^2}{a^2}\right) + \frac{dS}{S} = 0$$

or

$$\frac{du}{u}(N_{\text{Ma}}^2 - 1) = \frac{dS}{S}$$

3. Adiabatic Frictional Flow

a. Introduction

If heat transfer through the pipe wall is negligible, flow through a straight pipe of constant cross section is adiabatic.

b. Adiabatic Frictional Flow Equations

Property equations

$$\frac{p_a}{p_b} = \frac{N_{\text{Ma,b}}}{N_{\text{Ma,a}}}\sqrt{\left(1 + \frac{\gamma - 1}{2}N_{\text{Ma,b}}^2\right)\bigg/\left(1 + \frac{\gamma - 1}{2}N_{\text{Ma,a}}^2\right)}$$

$$\frac{T_a}{T_b} = \left(1 + \frac{\gamma - 1}{2}N_{\text{Ma,b}}^2\right)\left(1 + \frac{\gamma - 1}{2}N_{\text{Ma,a}}^2\right)$$

and

$$\frac{\rho_a}{\rho_b} = \frac{p_a}{p_b}\frac{T_b}{T_a}$$

where a is the entrance and b is the exit.

c. Maximum Length of Conduit

The maximum value of length L, consistent with a given entrance Mach number, is given by

$$\frac{\bar{f}L_{max}}{r_H} = \frac{1}{\gamma}\left[\frac{1}{N^2_{Ma,a}} - 1 - \frac{\gamma+1}{2}\ln\frac{2(1+(\gamma-1)/2N^2_{Ma,a})}{N^2_{Ma,a}(\gamma+1)}\right]$$

where $\bar{f} = \frac{1}{2}(f_a + f_b)$.

d. Mass Velocity

$$G = \rho N_{Ma}\sqrt{\frac{g_c\gamma TR^2}{M}} = N_{Ma}\sqrt{\rho g_c\gamma p}$$

4. Isothermal Frictional Flow

a. Introduction

The fluid temperature during compressible flow through a pipe of constant cross section can be kept constant by heat transfer through the pipe wall. The basic equation is

$$\frac{M}{2RT}(p_a^2 - p_b^2) - \frac{G^2}{g_c}\ln\frac{\rho_a}{\rho_b} = \frac{G^2 f\Delta L}{2g_c r_H}$$

where $\Delta L = L_b - L_a$.

b. Heat Transfer in Isothermal Flow

$$\frac{Q}{m} = \frac{G^2}{2g_c J}\left(\frac{1}{\rho_b^2} - \frac{1}{\rho_a^2}\right)$$

IV. FLOW PAST IMMERSED BODIES

A. Drag

Drag is the force in the direction of flow exerted by a fluid on a solid. When the solid's wall is parallel to the flow direction, the only drag force is the wall shear τ_w. The fluid pressure may possess a component in the direction of flow, and this also contributes to drag. The total drag is the sum of the two. Wall drag is the total integrated drag from wall shear; form drag is the total integrated drag from pressure. The phenomena in actual fluids are complicated; experimental or approximate methods are usually employed.

B.   Drag Coefficients

A drag coefficient is used for immersed solids, analogous to a friction factor. Just as the friction factor is defined as the ratio of τ_w to the product of fluid density and the velocity head, the drag coefficient C_D is defined as the ratio of F_D/A_P to this same product,

$$C_D = \frac{F_D/A_P}{\rho u_0^2/2g_c}$$

where F_D = total drag,
     A_P = projected area; area perpendicular to flow, and
     u_0 = velocity of approaching stream.

Using dimensional analysis, you can determine that the drag coefficient of a smooth solid in an incompressible fluid depends upon a Reynolds number and shape factors. For a given shape

$$C_D = \phi(N_{Re,p})$$

with $N_{Re,p} = G_0 D_P/\mu$,
     D_P = characteristic length and
     $G_0 = u_0\rho$.

C.   Particle Motion through Fluids

1.   Mechanics

Particle movement through a fluid requires a density difference between the fluid and the particles. In addition, an external force must provide motion to the particle relative to the fluid. This force is usually gravity. The forces acting on a particle that is moving through a fluid include

- an external force: gravity or centrifugal,
- a buoyant force acting parallel to the external force but in opposite direction, and
- a drag force acting to oppose the motion.

2. Equations for Particle Motion through a Fluid

A force balance gives

$$\frac{m}{g_c}\frac{du}{dt} = F_e - F_b - F_D$$

with $F_e = \dfrac{ma_e}{g_c}$ (external force)

                   a_e = acceleration of the particle

$F_b = \dfrac{m\rho a_e}{\rho_p g_c}$ (buoyant force)

                   ρ_p = particle density

$F_D = \dfrac{C_D u^2 \rho A_p}{2g_c}$ (drag force)

Then

$$\frac{du}{dt} = a_e \frac{\rho_D - \rho}{\rho_p} - \frac{C_D u^2 \rho A_p}{2m}$$

If the external force is gravity

$$a_e = g \text{ (the acceleration of gravity)}$$

If the external force is centrifugal

$$a_e = r\omega^2$$

where r = particle path radius and
        ω = angular velocity (radians/second).

3. Terminal Velocity

The particle quickly reaches a constant velocity, the maximum attainable if gravity is the force, and this is called the terminal velocity. We obtain it from $du/dt = 0$, or

$$u_t = \sqrt{\frac{2g(\rho_p - \rho)m}{A_p \rho \rho_p C_D}}$$

4. Drag Coefficients

The use of the above equations requires that C_D be known. When a particle is at sufficient distance from the container boundaries and other particles, we have free settling. If particle motion is impeded by other particles, we have hindered settling.

a. Spherical Particle Motion

Assume a sphere of diameter D_p, then

$$m = \frac{\pi D_p^3 \rho_p}{6}, \qquad A_p = \frac{\pi D_p^2}{4}$$

and

$$\frac{du}{dt} = a_e \frac{\rho_p - \rho}{\rho_p} - \frac{3 C_D u^2 \rho}{4 \rho_p D_p}$$

also

$$a_e(\rho_p - \rho) = 3 C_D u_f^2 \frac{\rho}{4 D_p}$$

at steady state.

b. Drag Coefficients for Spheres—Approximate

Stokes Law Range: $N_{Re,p} < 2$

$$C_D = \frac{24}{N_{Re,p}}, \qquad F_D = \frac{3 \pi \mu u_t D_p}{g_c}$$

Intermediate Range: $2 < N_{Re,p} < 500$

$$C_D = \frac{18.5}{N_{Re,p}^{0.6}}, \qquad F_D = \frac{2.31 \pi (u_t D_p)^{1.4} \mu^{0.6} \rho^{0.4}}{g_c}$$

Newton's Law Range: $500 < N_{Re,p} < 200{,}000$

$$C_D = 0.44, \qquad F_D = \frac{0.055 \pi (u_t D_p)^2 \rho}{g_c}$$

To use, calculate K from

$$K = D_p \left[\frac{a_e \rho(\rho_p - \rho)}{u^2} \right]^{1/3}$$

If	$K < 3.3$	use Stokes law
	$3.3 < K < 43.6$	Intermediate law
	$43.6 < K < 2360$	Newton's law

V. METERING AND TRANSPORTING FLUIDS

A. Fluid-Moving Machinery

1. Pumps

a. Developed Head

The Bernoulli equation can be used. The friction term h_f can be neglected since the only friction is that occurring in the pump (and this is taken into account with the mechanical efficiency η),

$$\eta W_p = \left(\frac{P_b}{\rho} + \frac{gZ_b}{g_c} + \frac{\alpha_b \bar{V}_b^2}{2g_c}\right) - \left(\frac{P_a}{\rho} + \frac{gZ_a}{g_c} + \frac{\alpha_a \bar{V}_a^2}{2g_c}\right)$$

The parentheses are the total heads H. Since a is the pump inlet and b the pump outlet, usually $Z_a \cong Z_b$. In any case H_a is the total suction head and H_b the total discharge head, or

$$W_p = \frac{H_b - H_a}{\eta} = \frac{\Delta H}{\eta}$$

The power supplied to the pump driver by some external source is P_B.

$$P_B = \dot{m} W_p = \frac{\dot{m}\Delta H}{\eta}, \qquad \dot{m} = \text{mass flow rate}$$

The power delivered to the fluid P_f is defined by $P_f/P_B = \eta$. These equations could also be used for fans by letting $\rho = (\rho_a + \rho_b)/2$.

The power is independent of pressure level, depending only on pressure difference. But the lower limit of the suction pressure is fixed by the vapor pressure at the temperature of the suction. If the suction pressure is lowered, some liquid flashes to vapor (called cavitation) and no liquid can be drawn into the pump. Of course, this will not occur if the sum of the velocity and pressure heads at the suction is greater than the liquid vapor pressure. The excess is called the net positive suction head (NPSH) or H_{sv}.

$$H_{sv} = \frac{\alpha_a \bar{V}_a^2}{2g_c} + \frac{p_a - p_v}{\rho}, \qquad p_v = \text{vapor pressure}$$

If the reservoir is located at a' below the pump entrance at a with friction h_{fs} in the suction line,

$$H_{sv} = \frac{p_a' - p_v}{\rho} - h_{fs} - \frac{gZ_a}{g_c}$$

where Z_a is the height of the pump above the reservoir.

b. Positive-Displacement Pumps

A definite volume of liquid is trapped in a chamber, alternately filled from the inlet and emptied through the discharge at a higher pressure. The two main types are reciprocating and rotary.

c. Centrifugal Pumps

The basic equations for power, developed head, and capacity have been derived for an ideal pump in McCabe and Smith, pp. 187–193. Their results are

$$\Delta H_r = H_b - H_a = \frac{u_2[u_2 - (q_r/A_p \tan \beta_2)]}{g_c}$$

where u_2 = peripheral velocity of tips of blades,

q_r = volumetric flow rate, and

β_2 = angle that tangent to impeller tip makes with tangent of circle traced out by impeller tips.

2. Compressors and Blowers

The general equation is

$$W_p = \int_{p_a}^{p_b} \frac{dp}{\rho}$$

a. Adiabatic Compression

If the unit is uncooled, the process is isentropic, so

$$W_p = \frac{p_a \gamma}{(\gamma - 1)\rho_a}\left[\left(\frac{p_b}{p_a}\right)^{(\gamma-1)/\gamma} - 1\right]$$

b. Isothermal Compression

If the unit is cooled, the process is isothermal, so

$$W_p = \frac{p_a}{\rho_a} \ln \frac{p_b}{p_a}$$

c. Polytropic Compression

Rather than being isothermal or isentropic, the machine is likely somewhere in between. It is useful to use

$$\frac{p}{\rho^n} = \frac{p_a}{\rho_a^n}$$

with n somewhere between 1 and γ. The constant n is determined experimentally from

$$n = \frac{\ln (p_b/p_a)}{\ln (\rho_b/\rho_a)}$$

Use this value n in place of the γ for the adiabatic compression.

d.   Power

The power for adiabatic compression or polytropic compression can be calculated from

$$P_B = \frac{0.0643 T_a \gamma q_0}{520(\gamma - 1)\eta} \left[\left(\frac{p_b}{p_a}\right)^{(\gamma - 1)/\gamma} - 1 \right]$$

where P_B = brake horsepower and
$\quad q_0$ = volume of gas compressed (std ft^3/min).

(Use n in place of γ if polytropic.)
    The power for isothermal compression is

$$P_B = \frac{0.248 T_a q_0}{520\eta} \log_{10}\left(\frac{p_b}{p_a}\right)$$

B.   Measurement of Flowing Fluids

1.   Venturi Meter

Using the Bernoulli equation for incompressible fluids and the continuity equation written as

$$\bar{V}_a = (D_b/D_a)^2 \bar{V}_b = \beta^2 \bar{V}_b, \qquad \beta = D_b/D_a$$

we obtain

$$\bar{V}_b = \frac{1}{\sqrt{\alpha_b - \beta^4 \alpha_a}} \sqrt{\frac{2g_c(p_a - p_b)}{\rho}}$$

for the Venturi meter shown in Figure F7. This is ideal. To account for the friction loss, the Venturi coefficient C_v is used:

$$\bar{V}_b = \frac{C_v}{\sqrt{1 - \beta^4}} \sqrt{\frac{2g_c(p_a - p_b)}{\rho}}$$

$$
\begin{array}{lll}
C_v \cong 0.98 & \text{for} & D_a \text{ 2-8 in.} \\
\cong 0.99 & \text{for} & D_a > 8 \text{ in.}
\end{array}
$$

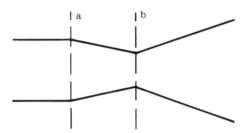

FIGURE F7. Venturi meter.

2. Orifice Meter

The same formula is appropriate for an orifice meter. If the Reynolds number, based on orifice diameter, is greater than 20,000, $C_v = 0.61$. In fact, if $\beta < 0.25$, the equation could be written

$$u_0 = 0.61 \sqrt{\frac{2g_c(p_a - p_b)}{\rho}}$$

3. Compressible Fluids with Venturi or Orifice Meters

For compressible fluids, multiply both formulas by Y. For Venturi meters,

$$Y = \left(\frac{p_b}{p_a}\right)^{1/\gamma} \sqrt{\frac{\gamma(1 - \beta^4)[1 - (p_b/p_a)^{(\gamma - 1)/\gamma}]}{(\gamma - 1)[1 - (p_b/p_a)][1 - \beta^4(p_b/p_a)^{2/\gamma}]}}$$

For orifice meters

$$Y = 1 - \frac{0.41 + 0.35\beta^4}{\gamma}\left(1 - \frac{p_b}{p_a}\right)$$

4. Pitot Tube

For incompressible fluids

$$u_0 = \sqrt{\frac{2g_c(p_s - p_0)}{\rho}}$$

and for compressible fluids

$$u_0 = \frac{g_c(p_s - p_0)}{\rho_0\left[1 + \dfrac{N_{Ma}^2}{4} + \dfrac{2 - \gamma}{24}N_{Ma}^4 + \cdots\right]}$$

using the notation of Figure F8.

5. Rotameters

These must be calibrated for a given fluid.

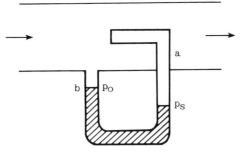

FIGURE F8. Pitot tube.

PROBLEMS

A. What is the settling time for the following spherical particles to settle under free settling conditions through 5 ft of water at 70°F. (From McCabe and Smith.)

Particle	Material	Specific Gravity	Diameter (in.)
1	Galena	7.5	0.01
2	Galena	7.5	0.001
3	Quartz	2.65	0.01
4	Quartz	2.65	0.001

B. A 10-ft diameter tank with a quick-opening value to a sump has a 10-ft depth of liquid. How long does it take to drain the tank if the valve is assumed to be equivalent to a 4-in. diameter orifice? The liquid density is 60 lb/ft³.

C. What size sharp-edged orifice was in use if a flow of 120 gpm produced a 4-in. Hg differential in a 4-in. Schedule 40 pipe. The fluid, measured at 60°F, had a viscosity of 1 centistoke and a specific gravity of 1.

D. Seven hundred gallons per minute of water are being discharged by a centrifugal pump 5 ft above the pump inlet. The inlet pressure is 4 in. Hg above atmospheric and the discharge pressure is 29 psia. If the pump has an 8-in. diameter inlet and a 4-in. diameter discharge, find (a) the pump efficiency, (b) the new flow rate, net head, and brake horsepower if the pump speed is increased from 1600 to 3200 revolutions per minute (rpm). Assume the pump input is 8 hp.

E. In case of a temperature runaway in our reactor (shown in Figure F9), cooling water (60°F) flows from a cooling tank (vented to the atmosphere, 14.7 psia) through the reactor to a waste system kept at 7 psia. The piping

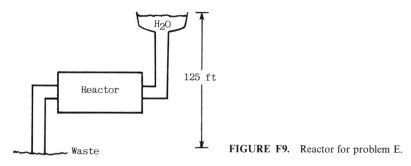

FIGURE F9. Reactor for problem E.

is 6-in. Schedule 40 steel containing two gate valves, three standard 90°
elbows, and three couplings and it has a length of 400 ft. Assume the pressure loss through the reactor is equivalent to another 500 ft of this pipe.
What will be the rate of water flow in an emergency (gpm)?

F.   A centrifuge with a 20-in. diameter bowl 14 in. deep operates at 1200 rpm
     giving a 2-in. thick liquid layer when used with a sludge whose liquid has a
     specific gravity of 1.2 and a viscosity of 2.5 CP. The solid component of the
     sludge has a specific gravity of 1.7. If the cut size of the particles is 35
     microns, what is the capacity of the centrifuge in gallons per minute?

G.   Determine the net positive suction head available to a centrifugal pump
     when the level of water is just at the suction point. The pumped fluid is
     water at 140°F and it is being pumped from an atmospheric tank at
     250 gpm. The tank suction point is 7 ft above the pump. The piping is 5-in.
     Schedule 40, 10-ft long with two short radius 90° elbows.

H.   Water at 60°F flows through a 3-in. Schedule 40 pipe. A pitot tube in the
     pipe shows a 3-in. Hg differential. If the pitot tube is located at the center of
     the pipe, what is the mass flow rate of the water?

PROBLEM-SOLVING STRATEGIES

A.   This is a free settling problem. How does the drag on a spherical particle
     affect settling time? Look up free settling or drag in the index of McCabe
     and Smith. Ahah! The equations are given. Just plug the data into the
     equations. A simple problem if the equations are available.

B.   Draining a tank is an unsteady-state process. Possibly Bird, Stewart, and
     Lightfoot[†] have the equation. If you cannot find the equation, develop one

---

[†] R. B. Bird, W. E. Stewart, and E. N. Lightfoot, *Transport Phenomena*, 1st. ed., Wiley, New-York,
1960.

using a force balance. An orifice equation relates force to pressure drop across the orifice. And the pressure drop is proportional to the height of fluid in the tank. We should be able to set up and solve the differential equation.

C. This orifice problem has all the data needed for a solution. Usually the size is given and the flow requested. Our problem will be trial and error since we will need to start by assuming that the orifice coefficient C_D is 0.61. With this assumption we use the orifice equation to determine the diameter. With the diameter, we calculate the Reynolds number, which allows us to check the value of C_D.

D. We can determine the shaft work with the Bernoulli equation since all the needed quantities are given. Since the pump input is given, we find the efficiency by dividing the shaft work by the pump input—using correct units of course. To solve the remainder of the problem, we need to know how pump speed affects the variables. Perry's *Chemical Engineers' Handbook* is the most likely source. If you cannot find it discussed, you should not attempt the problem unless you are willing to settle for half credit.

E. This is a Bernoulli equation problem with pipe friction. We know all quantities needed except the outlet velocity and the pipe friction. But the velocity and the pipe friction are related. So this must be a trial-and-error solution. One way to solve it is to assume a zero velocity and calculate the pipe friction h_f from the Bernoulli equation. Knowing h_f, we use the friction factor chart to find the Reynolds number and then the outlet velocity. Use this velocity in the Bernoulli equation to recalculate h_f. Then recalculate outlet velocity, and so forth. Except for the trial and error, the problem is simple.

F. This is a centrifuge problem and I know nothing about centrifuge design. To even start working this problem, I need an equation. Checking the McCabe and Smith index leads me in a short time to a centrifugal sedimentation design equation. All of the data needed for the equation is given in the problem. I have discovered that a seemingly difficult problem is a snap.

G. This problem can be solved with the pump Bernoulli equation. It looks like all needed data are given. Because the volumetric rate is given, no trial and error will be required. If I can find the definition of net positive suction head, I should have no difficulty with the solution.

H. This is a pitot tube problem and they are usually simple. I may need a correction since the velocity is a function of radius. A quick review of pitot tube equations should allow me to decide if I can handle the problem.

SOLUTIONS

A. From McCabe and Smith (McC&S), pp. 153–156 on free settling:

$$K = D_p Q \text{ where } Q = \left[\frac{a_e \rho(\rho_p - \rho)}{\mu^2}\right]^{1/3}, \quad a_e = g \qquad \text{(Eq. (7-54), McC\&S)}$$

If $\quad$ $K < 3.3$ $\qquad$ Stokes region,
$\quad$ $3.3 < K < 43.6$ $\qquad$ Intermediate,
$\quad$ $43.6 < K < 2360$ $\qquad$ Newton's law,

If Stokes,

$$u_t = \frac{a_e D_p^2(\rho_p - \rho)}{18\mu} \qquad \text{[Eq. (7-50), McC\&S]}$$

If intermediate,

$$u_t = \frac{0.153 a_e^{0.71} D_p^{1.14}(\rho_p \rho)^{0.71}}{\rho^{0.29}\mu^{0.43}} \qquad \text{[Eq. (7-51), McC\&S]}$$

Particle	Material	$\rho_p - \rho$	$g\rho/\mu^2$	Q	D_p(ft)	K	Region
1	Galena	405	4.61×10^9	12.33×10^3	0.000833	10.28	Intermediate
2	Galena	405	4.61×10^9	12.33×10^3	0.0000833	1.03	Stokes
3	Quartz	102.9	4.61×10^9	7.79×10^3	0.000833	6.49	Intermediate
4	Quartz	102.9	4.61×10^9		0.0000833	0.65	Stokes

where

$$\frac{g\rho}{\mu^2} = \frac{32.2 \text{ ft}}{\text{sec}^2} \frac{62.3 \text{ lb}}{\text{ft}^3} \frac{10^8 \text{ ft}^2 \text{ sec}^2}{6.60^2 \text{ lb}^2} = 4.61 \times 10^9$$

1. $u_t = \dfrac{(0.153)32.2^{0.71}(10^4)^{0.43}(0.000833^{1.14})405^{0.71}}{(62.3^{0.29})6.60^{0.43}} = 0.273 \dfrac{\text{ft}}{\text{sec}}$

$$t = \frac{5 \text{ ft}}{0.273 \text{ ft/sec}} = 18.3 \text{ sec}$$

2. $u_t = \dfrac{(32.2)8.33^2 \times 10^{-10}(10^4)}{(18)6.60}(405) = 0.0076 \dfrac{\text{ft}}{\text{sec}}$

$$t = 5/0.0076 = 658 \text{ sec}$$

3. $u_t = 0.1034$ ft/sec, $\qquad$ $t = 5/0.1034 = 48.3$ sec

4. $u_t = 0.001933$ ft/sec, $\qquad$ $t = 5/0.001933 = 2580$ sec

B. Consider a tank of cross-sectional area A and liquid height h as shown in Figure F10 and do a momentum balance with

$$\text{input} = 0, \qquad \text{output} = F, \qquad \text{accumulation} = A(dh/dt)$$

Then input−output = accumulation or

$$A\frac{dh}{dt} = -F$$

Now we know that the equation for an equivalent orifice is

$$F = C_D A_0 \sqrt{\frac{2g\Delta p}{\rho}}$$

and $C_D = 0.61$ and $\Delta p/\rho = h$. So

$$A(dh/dt) = -C_D A_0\sqrt{2gh}$$

or

$$\int h^{-1/2}\,dh = -\frac{C_D A_0\sqrt{2g}}{A}\int dt$$

or

$$2h^{1/2} = a - (C_D A_0\sqrt{2g}\,t/A)$$

At $t = 0$, $h = h_i = 10$ ft or $a = 2h_i^{1/2}$. So

$$t = \frac{2(h_i^{1/2} - h^{1/2})}{C_D A_0\sqrt{2g}}$$

$$A = \frac{\pi D^2}{4} = \frac{\pi}{4}(10\text{ ft})^2; \qquad A_0 = \frac{\pi}{4}\left(\frac{4\text{ in.}}{12\text{ in./ft}}\right)^2 = \frac{\pi}{4}\frac{1}{9}\text{ ft}^2$$

$$2g = 32.2 \times 2\text{ ft/sec}^2$$

When empty, $h = 0$

$$t = \frac{2A}{C_D A_0}\sqrt{\frac{h_i}{2g}} = \frac{2\pi 100\text{ ft}^2(4)(9)}{4(0.61)\pi\text{ ft}^2}\sqrt{\frac{10\text{ ft}}{64.4\text{ ft/sec}^2}}$$

$$= 1163\text{ sec} \quad\text{or}\quad 19.4\text{ min}$$

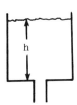

FIGURE F10. Schematic for solution B.

C.   We use McC&S Eqs. 8-39 to 8-43, following Example 8-5.

$$\rho = 62.3 \text{ lb/ft}^3, \qquad \Delta p = \frac{g \text{ lb}_f}{g_c \text{ lb}} \frac{4 \text{ in.}}{12 \text{ in.}}(13.6 - 1)62.3 \frac{\text{lb}}{\text{ft}^3} = 261.7 \frac{\text{lb}_f}{\text{ft}^2}$$

$$\dot{m} = \frac{120 \text{ gal}}{\text{min}} \frac{1 \text{ ft}^3}{7.48 \text{ gal}} \frac{62.3 \text{ lb}}{\text{ft}^3} \frac{1 \text{ min}}{60 \text{ sec}} = 16.66 \frac{\text{lb}}{\text{sec}}$$

4 in. Schedule 40 = 4.026 in. ID

$$D_a = \frac{4.026}{12} = 0.3355 \text{ ft}, \qquad g_c = 32.2 \frac{\text{lb ft}}{\text{sec}^2 \text{ lb}_f}$$

$$\beta^2 = \frac{4(16.66 \text{ lb/sec})}{0.61\pi(0.3355 \text{ ft})^2\sqrt{2(32.2 \text{ lb ft/sec}^2 \text{ lb}_f)[261.7 \text{ lb}_f/\text{ft}^2)(62.3 \text{ lb/ft}^3)}}$$

$$= 0.3015$$

$$\beta = 0.5491$$

So $D_0 = 0.5491(0.3355 \text{ ft}) = 0.184 \text{ ft} = 2.21$-in. orifice diameter. We must now check to see that the Reynolds number is $> 20,000$ in order to have used $C_D = 0.61$

$$\frac{\mu}{\rho} = 1 \text{ cS} \times 10^{-6} \frac{\text{m}^2/\text{sec}}{\text{cS}} = 10^{-6} \frac{\text{m}^2}{\text{sec}}$$

$$= 10^{-6} \frac{\text{m}^2}{\text{sec}} \left(\frac{3.2808 \text{ ft}}{\text{m}}\right)^2 = 10.76 \times 10^{-6} \frac{\text{ft}^2}{\text{sec}}$$

$$\mu = 10.76 \times 10^{-6}(62.3) \text{ lb/ft sec}$$

$$N_{Re,0} = \frac{4\dot{m}}{\pi D_0 \mu} = \frac{4(16.66 \text{ lb/sec})}{\pi(0.184 \text{ ft})(62.3)(10.76 \times 10^{-6}) \text{ lb/ft sec}}$$

$$= 177,976$$

So $C_0 = 0.61$ and our calculation is correct.

D.   Flow rate:

$$\frac{700 \text{ gal}}{\text{min}} \frac{1 \text{ ft}^3}{7.48 \text{ gal}} \frac{1 \text{ min}}{60 \text{ sec}} = 1.560 \frac{\text{ft}^3}{\text{sec}}$$

or

$$1.5560 \frac{\text{ft}^3}{\text{sec}} \times 62.3 \frac{\text{lb}}{\text{ft}^3} = 97.2 \frac{\text{lb}}{\text{sec}}$$

Use McC&S, Eq. 8-2.

(a) $$\eta W_p = \frac{P_b - P_a}{\rho} + \frac{g}{g_c}(Z_b - Z_a) + \frac{\alpha_b V_b^2 - \alpha_a V_a^2}{2g_c} \qquad \begin{array}{l} a = \text{inlet} \\ b = \text{outlet} \end{array}$$

$$W_P = 8 \text{ hp} \times \frac{550 \text{ ft lb}_f}{\text{sec}} \frac{\text{sec}}{1 \text{ hp}} \frac{1}{97.2 \text{ lb}} = 45.3 \frac{\text{ft lb}_f}{\text{lb}}$$

$$P_a = 4 \text{ in. Hg} + 29.92 \text{ in. Hg} = 33.92 \text{ in. Hg} \times \frac{14.7 \text{ psia}}{29.92 \text{ in. Hg}}$$

$$= 16.7 \text{ psia}$$

$$\bar{V}_a = 1.560 \frac{\text{ft}^3}{\text{sec}} \frac{4}{\pi \, 8^2 \text{ in.}^2} \frac{144 \text{ in.}^2}{\text{ft}^2} = 4.47 \frac{\text{ft}}{\text{sec}}$$

Now

$$\bar{V}_a A_a = \bar{V}_b A_b \qquad \text{so} \qquad \bar{V}_b = \frac{A_a}{A_b} \bar{V}_a = \frac{8^2}{4^2} \bar{V}_a = 4 \bar{V}_a$$

$$\bar{V}_b = 4(4.47 \text{ ft/sec}) = 17.88 \text{ ft/sec}$$

If we assume kinetic energy correction factors, α_a and $\alpha_b = 1$

$$\eta \left(45.3 \frac{\text{ft lb}_f}{\text{lb}} \right) = \frac{(29 - 17.2) \text{ lb}_f \text{ ft}^3}{\text{in.}^2 \, 62.3 \text{ lb}} \frac{144 \text{ in.}^2}{\text{ft}^2} \frac{5 \text{ lb}_f \text{ ft}}{\text{lb}}$$

$$= \frac{(17.88^2 - 4.47^2) \text{ ft}^2}{2(32.2 \text{ lb ft/sec}^2 \text{ lb}_f) \text{ sec}^2}$$

or

$$45.3\eta \frac{\text{ft lb}_f}{\text{lb}} = 27.3 \frac{\text{ft lb}_f}{\text{lb}} + 5 \frac{\text{ft lb}_f}{\text{lb}} + 4.7 \frac{\text{ft lb}_f}{\text{lb}} = 37 \frac{\text{ft lb}_f}{\text{lb}}$$

So $\eta = 0.82$ or efficiency $= 82\%$.

(b) Capacity is proportional to speed. Head is proportional to (speed)^2. Brake horsepower proportional to (speed)^3. So if speed is doubled, at the new conditions

$$\text{gal/min} = 2(700) = 1400 \text{ gal/min}$$
$$\text{head} = 4(37) = 148 \text{ ft lb}_f/\text{lb}$$
$$\text{bhp} = 8(8 \text{ hp}) = 64 \text{ hp}$$

E. We use the Bernoulli equation corrected for fluid friction, McC&S, Eq. (4-30).

$$\frac{p_a}{\rho} + \frac{gZ_a}{g_c} + \frac{\alpha_a V_a^2}{2g_c} = \frac{p_b}{\rho} + \frac{gZ_b}{g_c} + \frac{\alpha_b V_b^2}{2g_c} + h_f \qquad \begin{array}{l} a = \text{inlet (tank)} \\ b = \text{outlet (waste)} \end{array}$$

0 (at rest in tank)   0 (baseline)

We know neither V_b nor h_f; to determine h_f, we need $\bar{V}_b$. If we know h_f, we can calculate $\bar{V}_b$ from

$$N_{\text{Re}} \sqrt{f} = \frac{D\rho}{\mu} \sqrt{\frac{h_f D g_c}{2\Delta L}} \qquad \text{[Eq. (5-63), or Fig. 5-10, McC&S]}$$

As a first approximation, assume $\bar{V}_b = 0$; then

$$\frac{p_a - p_b}{\rho} + \frac{gZ_a}{g_c} = \frac{\alpha_b \bar{V}_b^2}{2g_c} + h_f$$

or

$$\frac{(14.7 - 7)\ lb_f\ ft^3}{in.^2\ 62.3\ lb}\frac{144\ in.^2}{ft^2} + 125\ \frac{ft\ lb_f}{lb} = \frac{\alpha_b \bar{V}_b^2}{2g_c} + h_f$$

$$\text{0, 1st approximation}$$

or

$$h_f = 17.8 + 125 = 142.8\ ft\ lb_f/lb$$

Then

$$N_{Re}\sqrt{f} = \frac{6.065\ in.}{12\ in./ft}\frac{62.4\ lb\ ft\ sec}{ft^3\ 7.59 \times 10^{-4}\ lb}$$

$$\times \sqrt{\frac{142.8\ ft\ lb_f\ 6.065\ ft\ 32.2\ ft\ lb}{lb\ (2)900\ ft\ 12\ sec^2\ lb_f}}$$

$$= 4.155 \times 10^4\sqrt{1.291} = 4.72 \times 10^4$$

Using Fig. 5-10 with $K = 0.00015$,

$$\frac{K}{D} = \frac{0.00015}{6.065} \times 12 = 0.0003 \qquad \text{or} \qquad 0.004 = f$$

So

$$N_{Re} = \frac{4.72 \times 10^4}{\sqrt{f}} = \frac{4.72 \times 10^4}{\sqrt{0.004}} = 74.63 \times 10^4$$

$$N_{Re} = \frac{DV\rho}{\mu} \qquad \text{or} \qquad V = \frac{\mu}{D\rho} \times 74.63 \times 10^4$$

$$\bar{V}_b = \frac{7.59 \times 10^{-4}\ lb\ 12\ ft^3}{ft\ sec\ 6.065\ ft\ 62.4\ lb} \times 74.63 \times 10^4 = 17.96\ \frac{ft}{sec}$$

We now check this in the Bernoulli equation, using

$$\left(4f\frac{L}{D} + K_f\right)\frac{\bar{V}_b^2}{2g_c}$$

in place of h_f (see McC&S, Eq. 5-74).
   So

$$142.8\ \frac{ft\ lb_f}{lb} = \frac{\alpha_b \bar{V}_b^2}{2g_c} + \left(4f\frac{L}{D} + K_f\right)\frac{\bar{V}_b^2}{2g_c}$$

From Table 5-1, McC&S

$$K_f$$

$$\left.\begin{array}{l} \text{2 gate valves} = 2(0.2) \\ \text{3 } 90° \text{ elbows} = 3(0.9) \\ \text{3 couplings} = 3(0) \end{array}\right\} 3.1 \qquad \text{if} \quad \alpha_b = 1$$

$$142.8 = \left[1 + \frac{4(0.004)900(12)}{6.065} + 3.1\right]\frac{\bar{V}_b^2}{64.4} = 0.5061\bar{V}_b^2$$

$$\bar{V}_b = 16.80 \text{ ft/sec}$$

This value of $\bar{V}_b$ will not change the f significantly.

$$\text{inside area} = 0.2006 \text{ ft}^2$$

$$\text{gpm} = 16.8 \frac{\text{ft}}{\text{sec}}(0.2006 \text{ ft}^2)7.48\frac{\text{gal}}{\text{ft}^3}60\frac{\text{sec}}{\text{min}} = 1512 \text{ gpm}$$

F. See Figure F11. We use McC&S, Eq. 30-55.

$$q_c = \frac{\pi b \omega^2(\rho_p - \rho)D_{pc}^2}{18\mu}(r_2^2 - r_1^2)/\ln\left(\frac{2r_2}{r_1 + r_2}\right)$$

$$D_{pc} = 35 \text{ microns} \times \frac{3.973 \times 10^{-5} \text{ in.}}{\text{micron}} \times \frac{\text{ft}}{12 \text{ in.}}$$

$$= 1.148 \times 10^{-4} \text{ ft}$$

$$\rho_p - \rho = (1.7 - 1.2)62.4\frac{\text{lb}}{\text{ft}^3} = 31.2\frac{\text{lb}}{\text{ft}^3}$$

$$\omega = 1200 \text{ rpm}\frac{2\pi \text{ min}}{60 \text{ sec}} = 125.7\frac{\text{rad}}{\text{sec}}$$

$$\mu = 2.5 \text{ cp}\frac{6.7197}{\text{cp}} \times 10^{-4}\frac{\text{lb}}{\text{ft sec}} = 16.80 \times 10^{-4}\frac{\text{lb}}{\text{ft sec}}$$

$$r_2 = \frac{10 \text{ in.}}{12 \text{ in/ft}} = 0.833 \text{ ft}; \qquad r_1 = \frac{8 \text{ in.}}{12 \text{ in./ft}} = 0.666 \text{ ft}$$

$$b = \frac{14 \text{ in.}}{12 \text{ in./ft}} = 1.167 \text{ ft}$$

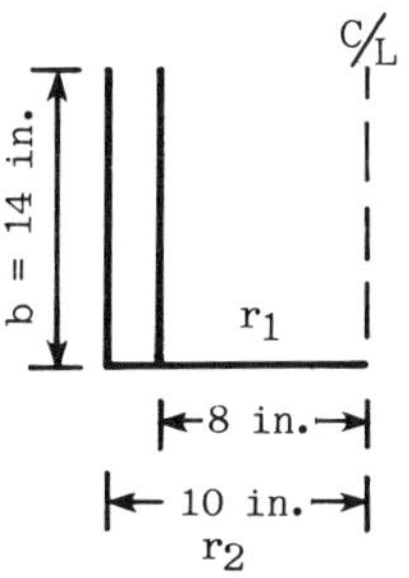

FIGURE F11. Schematic for solution F.

$$q_c = \frac{\pi(1.167 \text{ ft})125.7 \text{ rad}^2/\text{sec } 31.2 \text{ lb/ft}^3 (1.148 \times 10^{-4})^2}{18(16.80 \times 10^{-4} \text{ lb/ft sec})}$$

$$\cdot (0.833^2 - 666^2 \text{ ft}^2)/\ln\left[\frac{2(0.833)}{(0.833 + 0.666)}\right]$$

$$= 1.879 \frac{\text{ft}^3}{\text{sec}} \frac{7.48 \text{ gal}}{\text{ft}^3} \frac{60 \text{ sec}}{\text{min}} = 843 \frac{\text{gal}}{\text{min}}$$

G. See Figure F12. From McC&S, Eq. 8-8, VP at $150°F = p_v$.

$$H_{sv} = \frac{\alpha_a \bar{V}_a^2}{2g_c} + \frac{p_a - p_v}{\rho}$$

$$\text{vol rate} = \frac{250 \text{ gal ft}^3}{\text{min}} \frac{1 \text{ min}}{7.48 \text{ gal}} \frac{1 \text{ min}}{60 \text{ sec}} = 0.557 \frac{\text{ft}^3}{\text{sec}}$$

$$\bar{V}_a = \frac{0.557 \text{ ft}^3}{A_i \text{ sec}}, \qquad A_i = \text{cross-sectional area of 5 in. Schedule 40 pipe} =$$

$$0.139 \text{ ft}^2$$

$$\bar{V}_a = \frac{0.557 \text{ ft}^3}{0.139 \text{ ft}^2 \text{ sec}} = 4.01 \frac{\text{ft}}{\text{sec}}, \qquad D_i = \frac{5.047 \text{ in.}}{12 \text{ in. ft}} = 0.4206 \text{ ft}$$

So we know everything but p_a. We use Bernoulli with $\rho = 61.38 \text{ lb/ft}^3$, $p_v = 2.8886 \text{ lb}_f/\text{in.}^2$, and $\mu = 0.470 \text{ cP}$.

$$\frac{p_a'}{\rho} + \frac{gZ_a'}{g_c} + \frac{\alpha_a' \bar{V}_{a1}^2}{2g_c} = \frac{p_a}{\rho} + \frac{gZ_a}{g_c} + \frac{\alpha_a \bar{V}_a^2}{2g_c} + h_f$$

$$h_f = \left(4f \frac{L}{D} + K_f\right)\frac{\bar{V}_a^2}{2g_c}, \qquad K_f = 2(0.9) = 1.8 \qquad \text{McC\&S, Table 5-1}$$

$$N_{Re} = \frac{0.4206 \text{ ft}}{\text{sec}} \frac{4.01 \text{ ft}}{\text{ft}^3} \frac{61.38 \text{ lb}}{0.470 \text{ cp}} \frac{\text{cp ft sec}}{6.7197 \times 10^{-4} \text{ lb}} = 3.28 \times 10^5$$

$$\frac{k}{D} = \frac{0.00015}{0.4206} = 0.00036$$

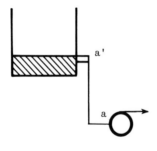

FIGURE F12. Schematic for solution G.

Figure 5-9 gives $f = 0.0043$, so

$$h_f = \left[4(0.0043) \frac{7}{0.4206} + 1.8 \right] \frac{(4.01 \text{ ft/sec})^2 \text{ sec}^2 \text{ lb}_f}{2(32.2) \text{ ft lb}} = 0.521 \frac{\text{ft lb}_f}{\text{lb}}$$

Now for the Bernoulli equation

$$\frac{\alpha_a \bar{V}_a^2}{2g_c} + \frac{p_a}{\rho} = \frac{p_a'}{\rho} + \frac{gZ_a'}{g_c} - h_f = H_{sv} + \frac{p_v}{\rho}$$

So

$$H_{sv} = \frac{p_a' - p_v}{\rho} + \frac{gZ_a'}{g_c} - h_f$$

$$= \frac{\left(14.7 \frac{\text{lb}_f}{\text{in.}^2} - 2.89 \frac{\text{lb}_f}{\text{in.}^2} \right) \frac{144 \text{ in.}^2}{\text{ft}^2}}{61.38 \text{ lb/ft}^3} + \frac{\text{lb}_f}{\text{lb}} 7 \text{ ft} - 0.521 \frac{\text{ft lb}_f}{\text{lb}}$$

$$= (27.7 + 7 - 0.5) \text{ ft lb}_f/\text{lb} = 34.2 \text{ ft lb}_f/\text{lb}$$

H. Using McC&S, Eq. 8-49

$$u_0 = \sqrt{\frac{2g_c \Delta p}{\rho}}, \qquad u_0 = \text{centerline velocity, max} = u_{max}$$

Then use Figure 5-7,

$$\frac{\bar{V}}{u_{max}} \Bigg|_{N_{Re}}$$

to obtain $\bar{V}$, with $\mu = 1.129 \text{ cP}$

$$p = \frac{g}{g_c} \frac{3 \text{ in.}}{12 \text{ in./ft}} (13.6 - 1)62.3 \frac{\text{lb}}{\text{ft}^3} = 196.2 \frac{\text{lb}_f}{\text{ft}^2}$$

$$u_0 = \sqrt{2 \frac{(32.2 \text{ ft lb})}{\text{sec}^2 \text{ lb}_f} \frac{196.2 \text{ lb}_f \text{ ft}^3}{\text{ft}^2 62.3 \text{ lb}}} = 14.24 \frac{\text{ft}}{\text{sec}}$$

$$N_{Re} = \frac{3.068 \text{ in.}}{12 \text{ in./ft}} 14.24 \frac{\text{ft}}{\text{sec}} \frac{62.3 \text{ lb}}{1.129 \text{ ft}^3} \frac{\text{ft sec}}{(6.7197 \times 10^{-4} \text{ lb})} = 3.38 \times 10^5$$

So

$$\bar{V}/u_{max} = 0.815$$

So

$$\bar{V} = 0.815(14.24 \text{ ft/sec}) = 11.61 \text{ ft/sec}$$

or

$$(11.61 \text{ ft/sec})0.0513 \text{ ft}^2(62.3 \text{ lb/ft}^3) = 37.1 \text{ lb/sec}$$

ENGINEERING ECONOMICS

E. L. Grant, W. G. Ireson, and R. S. Leavenworth, *Principles of Engineering Economy*, 7th ed., Wiley, New York, 1982.

H. G. Thuesen, W. J. Fabrycky, and G. J. Thuesen, *Engineering Economy*, 5th ed., Prentice-Hall, Englewood Cliffs, New Jersey, 1977.

I. INTEREST AND INTEREST FORMULAS

A. The Time Value of Money

Money can be invested so as to produce more money. If it is invested at some specific rate of interest it will increase in value with time. So money has a time value. With this in mind, the comparison of sums of money separated by time must be made at a common time.

B. Types of Interest

The rental rate for a sum of money is expressed as the percentage of the sum that is to be paid for the use of the sum for a period of one year. Interest rates are also given for periods other than one year. This will be discussed later.

1. Simple Interest

In simple interest, the interest to be paid on repayment of a loan is proportional to the length of time the principal sum has been borrowed. Let P represent the principal, n the interest period, and i the interest rate. Then the simple interest I is given by

$$I = Pni$$

2. Compound Interest

When a loan is made for a length of time equal to several interest periods, interest is calculated at the end of each interest period. A number of loan repayment plans are available. These range from paying the interest when it is due to accumulating all the interest until the loan is due. If the borrower does not pay the interest earned at the end of each period and is charged interest on the total amount owed (interest plus principal), the interest is said to be compounded.

C. The Cash Flow Diagram

To help in recording and identifying the economic effects of alternative investments, a graphical description of each alternative's cash transactions can be used. This pictorial description, called a cash flow diagram, will provide all the necessary information for analyzing an investment proposal.

The cash flow diagram represents any receipts received over a period of time as an upward arrow (a cash increase) located at the end of the period. The height of the arrow is usually made proportional to the magnitude of the receipts received during that period. Likewise, the disbursements are represented by a downward arrow (a cash decrease). These arrows are placed on a time scale that represents the duration of the proposal.

D.   Interest Formulas

The interest factors presented in this section apply to the common situation of annual compounding of interest and annual payments. We use the following symbols:

i = the annual interest rate,

n = the number of annual interest periods,

P = a present principal sum,

A = a single payment (annuity), in a series of n equal payments, made at the end of each annual interest period, and

F = a future sum, n annual interest periods hence, equal to the compound amount of a present principal sum P, or equal to the sum of the compound amounts of payments A, in a series.

The symbols are chosen so that each is an initial letter of a key word. Thus i applies to interest, n applies to number of periods, P applies to present worth (or sum), F applies to future worth, and A applies to annual payment or annuity.

1.   Single-Payment, Compound-Amount Factor

If an amount P is invested now with the amount earning at the rate i per year, how much principal and interest are accumulated after n years?

This is determined from the single-payment compound amount factor

$$F/P(i, n) = (1 + i)^n$$

or

$$F = P \cdot [F/P(i, n)]$$

as shown in Figure E1.

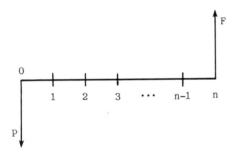

FIGURE E1.   Cash flow diagram—single-payment compound-amount.

2. *Single-Payment, Present-Worth Factor*

The single-payment, compound-amount relationship can be solved for P

$$P = F \frac{1}{(1 + i)^n} = F \cdot \left[\frac{P}{F(i, n)} \right]$$

The resulting factor is known as the single-payment, present-worth factor.

3. *Equal-Payment-Series, Compound-Amount Factor*

In engineering economics studies it is often necessary to find the single future value that would accumulate from a series of equal payments occurring at the end of succeeding annual interest periods. The factor needed is

$$F = A \left[\frac{(1 + i)^n - 1}{i} \right] = A \cdot \left[\frac{F}{A(i, n)} \right]$$

its cash flow diagram is illustrated in Figure E2.

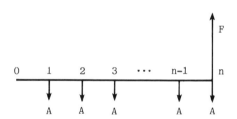

FIGURE E2. Cash flow diagram—series-payments compound-amount.

4. *Equal-Payment-Series, Sinking-Fund Factor*

The equal-payment-series, compound-amount relationship can be solved for A

$$A - F \left[\frac{i}{(1 + i)^n - 1} \right] = F \cdot \left[\frac{A}{F(i, n)} \right]$$

This is called the equal-payment-series, sinking-fund factor.

5. *Equal-Payment-Series, Capital-Recovery Factor*

A deposit of amount P is made now at an annual interest rate i. The depositor wants to withdraw the principal plus earned interest in a series of equal year-end amounts over the next n years. When the last withdrawal is made there should be no funds left on deposit, as shown in Figure E3.

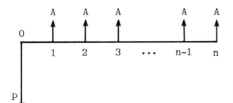

FIGURE E3. Cash flow diagram—series-payments capital-recovery.

The relationship needed is

$$A = P\left[\frac{i(1 + i)^n}{(1 + i)^n - 1}\right] = P \cdot \left[\frac{A}{P(i, n)}\right]$$

Note that for perpetual or infinite time, the capital-recovery factor is i.

6. Equal-Payment-Series, Present-Worth Factor

To determine what single amount must be deposited now so that equal end-of-year payments can be made, we solve for P in terms of A or

$$P = A\left[\frac{(1 + i)^n - 1}{i(1 + i)^n}\right] = A \cdot \left[\frac{P}{A(i, n)}\right]$$

Note that for perpetual or infinite time, the present-worth factor is $1/i$.

7. Uniform-Gradient-Series Factor

Frequently, annual payments do not occur in equal-payment series. A linearly increasing or decreasing series, as illustrated in Figure E4, can be reduced to an equivalent equal-payment series. If we let

$$A_1 = \text{payment at the end of the first year,}$$
$$G = \text{annual change or gradient, and}$$
$$A = A_1 + A_2,$$

then

$$A_2 = G\left[\frac{1}{i} - \frac{n}{i} \cdot \frac{A}{F(i, n)}\right] = G \cdot \left[\frac{A}{G(i, n)}\right]$$

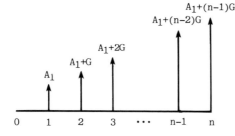

FIGURE E4. Cash flow diagram—uniform-gradient-series.

8. Format for the Use of Interest Factors

In engineering economics studies, disbursements made to start an alternative are considered to take place at the beginning of the period embraced by the alternative. Payments occurring during the period of the alternative are usually assumed to occur at the end of the year or interest period in which they occur.

In the use of interest factors for annual payments:

- The end of one year is the beginning of the next year.
- P is at the beginning of a year at a time regarded as being the present.
- F is at the end of the nth year, from a time regarded as being the present.

- A occurs at the end of each year. When P and A are involved, the first A of the series occurs one year after P. When F and A are involved, the last A of the series occurs at the same time as F.

9. Interest Factor Relationships

$$P/F(i, n) = 1/[F/P(i, n)]$$
$$F/A(i, n) = 1/[A/F(i, n)]$$
$$P/A(i, n) = 1/[A/P(i, n)]$$
$$F/A(i, n) = 1 + F/P(i, 1) + F/P(i, 2) + \cdots + F/P(i, n - 1)$$
$$P/A(i, n) = P/F(i, 1) + P/F(i, 2) + \cdots + P/F(i, n)$$
$$A/P(i, n) = A/F(i, n) + i$$
$$F/G(i, n) = [A/G(i, n)][F/A(i, n)]$$
$$F/P(i, n) = [F/P(i, n_1)][F/P(i, n_2)] \cdots [F/P(i, n_k)]$$
$$\text{where} \quad n = n_1 + n_2 \cdots + n_k$$

E. Nominal and Effective Interest Rates

Up to this point, we have looked at interest periods of only one year. But agreements can specify that interest will be paid more frequently, such as each half-year, each quarter, or each month. These agreements result in interest periods of one-half, one-quarter, or one-twelfth year, and the compounding of interest 2, 4, or 12 times a year, respectively.

The interest rates associated with this more frequent compounding are normally quoted on an annual basis according to the following convention. When the actual or effective rate of interest is 3% interest compounded each six-month period, the annual or nominal interest is quoted as 6% per year compounded semiannually. For an effective rate of interest of 1.5% compounded at the end of each three-month period the nominal interest is quoted as 6% per year compounded quarterly. So the nominal rate of interest is expressed on an annual basis and it is determined by multiplying the actual or effective interest rate per interest period times the number of interest periods per year. Thus the actual or effective interest rate per year is higher than the nominal interest rate. If we let

$$r = \text{nominal interest rate (per year)},$$
$$i = \text{the effective interest rate (per period)},$$
$$c = \text{the number of interest periods per year},$$

then

$$i = \text{effective annual interest rate} = (1 + r/c)^c - 1$$

The previous interest formulas were derived on the basis of an effective interest rate for an interest period, specifically, for an annual interest rate compounded annually. But they can be used when compounding occurs more frequently than once a year. This can be done in either of two ways:

1. find the effective annual interest rate from the previous formula and use this rate when time periods are years; or
2. match the interest rate to the interest period and use the factors directly or its corresponding tabulated value.

For example, we wish to find the compound amount of $1000 four years from now at a nominal annual interest rate of 6% compounded semiannually.
Since

$$c = 2 \quad \text{and} \quad r = 0.06$$

$$i = \left(1 + \frac{0.06}{2}\right)^2 - 1 = 0.0609$$

$$F = \$1000 \, (1 + 0.0609)^4 = \$1267$$

or

$$F = \$1000 \cdot F/P(6.09\%, 4) = \$1267$$

Likewise, since the nominal annual interest rate is 6% compounded semiannually, the interest rate is 3% for an interest period of one-half year. So

$$F = \$1000 \, (1 + 0.030)^8 = \$1267$$

or

$$F = \$1000 \cdot F/P(3\%, 8) = \$1267$$

II. BASES FOR COMPARISON OF ALTERNATIVES

All decision criteria must incorporate some index, measure of equivalence, or bases for comparison that summarizes the significant differences between investment alternatives. A basis for comparison is an index containing information about a series of receipts and disbursements representing an investment opportunity. This reduction of alternatives to a common base is needed so that apparent differences become real differences with the time value of money considered. These real differences become directly comparable and can be used in decision making. The most common bases are

- Present-worth amount.
- Annual equivalent amount.
- Capitalized equivalent amount.
- Future-worth amount.
- Rate of return.

A. Net Cash Flows for Investment Opportunities

An investment is described by the cash receipts and disbursements that are anticipated. The representation of the amounts and timing of these cash receipts and disbursements is called the cash flow of the investment.

When an investment has both cash receipts and disbursements occurring simultaneously, a net cash flow is calculated for the investment. The net cash flow is the arithmetic sum of the receipts $(+)$ and the disbursements $(-)$ that occur at the same point in time. This utilization of net cash flows in decision making implies that the net dollars received or disbursed have the same effect on an investment decision as does the separate consideration of the total receipts and disbursements of the investment.

To describe investment cash flows we shall use the notation

$$F_{jt} = \text{net cash flow for proposal } j \text{ at time } t$$

Then if $F_{jt} < 0$, F_{jt} represents a net cash cost or disbursement. If $F_{jt} > 0$, F_{jt} is a net cash income or receipt.

B. Present-Worth Amount

An index that reflects the differences between alternatives by considering the time value of money is the present-worth amount. For a known cash flow and given interest rate, an equivalent amount for the cash flow can be calculated at any point in time. So it is possible to calculate a single equivalent amount at any point in time that is equivalent in value to a particular cash flow.

The present-worth amount is an amount at the present $(t = 0)$ that is equivalent to the cash flow of an investment for a particular interest rate i. So the present worth of investment proposal j at interest rate i with a life of n years is

$$PW(i)_j = F_{j0} \cdot P/F(i, 0) + F_{j1} \cdot P/F(i, 1) + \cdots + F_{jn} \cdot P/F(i, n)$$

$$= \sum_{t=0}^{n} F_{jt} \cdot P/F(i, t)$$

A number of features makes the present-worth amount suitable as a basis for comparison:

- It considers the time value of money according to the value of i selected for calculations.
- It concentrates the equivalent value of any cash flow into a single index at a particular time ($t = 0$).
- The present-worth amount is unique regardless of the investment's cash flow pattern.
- The present-worth amount is the equivalent amount by which the equivalent receipts of a cash flow exceed (or fail to equal) the equivalent disbursements of that cash flow.

C. Capitalized Equivalent

A special case of the present-worth amount $PW(i)_j$ is the capitalized equivalent $CE(i)$. This method finds a single amount at the present which at a given rate of interest will be equivalent to the net difference of receipts and disbursements if a given cash flow pattern is repeated in perpetuity.

To determine the capitalized equivalent of an investment that is to produce cash flows from the present to infinity, we:

- Convert the actual cash flow into an equivalent cash flow of equal annual payments A that extends to infinity.
- Discount the equal annual payments A to the present by using the equal-payment-series, present-worth factor.

D. Annual Equivalent Amount

The annual equivalent amount has characteristics similar to the present-worth amount. Any cash flow can be converted into a series of equal annual payments by first calculating the present-worth amount for the original series and then multiplying the present-worth amount by the interest factor $A/P(i, n)$. So the annual equivalent amount $AE(i)$ for interest rate i and n years can be calculated from

$$AE(i) = PW(i) \cdot A/P(i, N)$$

E. Capital Recovery with Return

An asset is a unit of capital. Some assets, such as machines, lose value over a period of time. This loss of value of an asset represents a consumption or expenditure of capital.

Capital assets are bought with the idea that they will earn more than they cost. So a part of the prospective earnings is considered to be capital recovery. The capital is recovered in the form of income derived from the services rendered by the asset and from its sale at the end of its useful life.

Another part of the prospective earnings will be considered to be return. Since the capital invested in an asset is usually recovered bit by bit, it is necessary to consider the interest on the unrecovered balance as a cost of ownership.

Two transactions are associated with the buying and eventual retirement of a capital asset; its first cost and salvage value. A simple formula for the equivalent annual cost of an asset can be determined from these amounts. We let

$$P = \text{first cost of an asset,}$$
$$F = \text{estimated salvage value,}$$
$$n = \text{estimated service life in years, and}$$
$$CR(i) = \text{capital recovery with return.}$$

So the annual equivalent cost of an asset can be expressed as the annual equivalent first cost less the annual equivalent salvage value, or

$$CR(i) = P \cdot A/P(i, n) - F \cdot A/F(i, n)$$

or

$$CR(i) = (P - F) \cdot A/P(i, n) + Fi$$

F. Future-Worth Amount

The future-worth basis for comparison is an equivalent amount of a cash flow calculated at a future time for some interest rate. The future-worth amount for a proposal j at some future time n years from the present is given by

$$FW(i)_j = P_{j0} \cdot F/P(i, n) + P_{j1} \cdot F/P(i, n - 1) + \cdots + P_{jn} F/P(i, 0)$$

or

$$FW(i)_j = \sum_{t=0}^{n} P_{jt} \cdot F/P(i, n - t)$$

A simple method for calculating the future-worth amount is to first determine the present-worth amount of the cash flow and then to convert to its future equivalent n year hence, or

$$FW(i)_j = PW(i)_j \cdot F/P(i, n)$$

G.   Rate of Return

The rate of return is a widely accepted index of profitability. It is the interest rate that reduces the present-worth amount of a series of receipts and disbursements to zero. So, the rate of return for investment proposal j is the interest rate i_j^* that satisfies the equation

$$0 = PW(i_j^*)_j = \sum_{t=0}^{n} F_{jt} \cdot P/F(i_j^*, t)$$

Because of the relationship between the annual equivalent amount, the future-worth amount, and the present-worth amount, this equation is also equivalent to

$$0 = AE(i_j^*)_j$$

or

$$0 = FW(i_j^*)_j$$

In all the other methods, the interest rate is given (or assumed given). With this method, we must calculate the interest rate or rate of return. The calculation becomes a trial-and-error problem. But in return for the extra work, we have a reasonable way to compare the economic desirability of alternative investments without assuming that we know highly uncertain future interest rates.

H.   Some Relationships between Methods

Given one basis for comparison, we can calculate the others from the following relationships:

$$PW(i) = AE(i) \cdot P/A(i, n)$$

$$CE(i) = AE(i)/i$$

$$PW(i) = CE(i) \cdot i \cdot P/A(i, n)$$

$$FW(i) = PW(i) \cdot F/P(i, n)$$

III.   DECISION MAKING AMONG ALTERNATIVES

We now look at various decision criteria that describe how investment decisions should be made. These criteria will be rules or procedures that describe how to select investment opportunities so that particular objectives can be achieved.

A.   Types of Investment Proposals

An investment proposal is considered to be a single project that is being considered as an investment possibility. An investment alternative is a decision option, so every investment proposal could be considered to be an investment alternative. However, an investment alternative could consist of a group or set of investment proposals. It could also represent the option of doing nothing. So if there are two proposals, A and B, it is possible to have four decision alternatives: do nothing, accept only A, accept only B, or accept both A and B.

1. Independent Proposals

When the acceptance of a proposal from a set of proposals has no effect on the acceptance of any of the other proposals contained in the set, the proposal is said to be independent. Usually, if proposals are functionally different and there are no other obvious dependencies between them, it is reasonable to consider the proposals as independent.

2. Dependent Proposals

For many decision problems, a group of investment proposals will be related in such a way that the acceptance of one of these proposals will influence the acceptance of the others. There are several types.

a. Mutually Exclusive Proposals

If the proposals in the set are related so that acceptance of any proposal from the set precludes the acceptance of any of the other proposals in the set, the proposals are said to be mutually exclusive. These usually occur when there is a need to be fulfilled and there are a variety of proposals each of which will satisfy that need.

b. Contingent Proposals

Another type occurs once some initial project is undertaken. Then a number of other auxiliary investments become feasible. Such auxiliary proposals are called contingent proposals because their acceptance is conditional on the acceptance of another proposal.

B. Mutually Exclusive Alternatives and Decision Making

From now on the selection of investment proposals will be viewed as a problem of selecting a single economic alternative from a set of alternatives. Thus the proposals are set up so that the alternatives are considered to be mutually exclusive.

1. Comparing Mutually Exclusive Alternatives

When comparing mutually exclusive alternatives, it is the future difference between the alternatives that is relevant for determining the economic desirability of one compared to the other.

2. Forming Mutually Exclusive Alternatives

Investment proposals can be independent, mutually exclusive, or contingent. Rather than develop special rules for each, the approach will be to arrange all proposals so that the selection decision involves only the consideration of the cash flows of mutually exclusive alternatives.

All that is required is the enumeration of all the feasible combinations of the proposals under consideration. Each combination represents a mutually exclusive alternative since each combination precludes the acceptance of any of the other combinations. The cash flow of each alternative is found simply by adding, period by period, the cash flows of each proposal contained in the alternative being considered.

C. Decision Criteria for Mutually Exclusive Alternatives

1. Selecting an Interest Rate

The objective is the maximization of equivalent profit given that all investment alternatives must yield a return that exceeds some minimum attractive rate of return (MARR). This cut-off rate is usually the result of a policy decision made by management. The MARR could be viewed as a rate at which the business can always invest since it has a large number of opportunities that yield such a return. Whenever any money is committed to an investment proposal, an opportunity to invest that money at the MARR has been foregone.

2. The Do Nothing Alternative

If the funds available are not invested in the projects being considered they will be invested in the do nothing alternative. This means that the investor will do nothing about the projects being considered but the funds will be placed in investments that yield a rate of return equal to the MARR.

3. Criteria

a. Present-Worth on Total Investment

This criterion is one of the most used criteria for selecting an investment alternative from a set of mutually exclusive alternatives. All that is required is to calculate the present-worth amount for the cash flow representing each alternative. Then select the alternative that has the maximum present-worth amount (provided this amount is positive).

We have seen that the present-worth amount, the annual equivalent amount, and the future-worth amount are consistent bases for comparing alternatives. Thus, if

$$PW(i)_A > PW(i)_B$$

then

$$AE(i)_A > AE(i)_B$$

and

$$FW(i)_A > FW(i)_B$$

Note that the future-worth amounts indicate that the receipts from the investment are actually invested at the MARR from the time they are received to the end of the life of the alternative. So future-worth calculations explicitly consider the investment or reinvestment of the future receipts generated by investment alternatives. So the use of the present-worth amount implicitly assumes the investment or reinvestment of receipts at the MARR.

b. Present-Worth on Incremental Investment

The present-worth on incremental investment requires that the incremental differences between alternative cash flows actually be calculated. We first determine the cash flow representing the difference between the two cash flows. Then the decision whether to select a particular alternative rests on the determination of the economic desirability of the additional increment of investment required by one alternative over the other. If the present-worth amount for the incremental investment is greater than zero, the increment is considered desirable and the alternative requiring this additional investment is deemed best.

To apply this decision criteria to a set of mutually exclusive alternatives, the following steps must be used:

• List the alternatives in ascending order of their equivalent first cost or initial disbursements.
• Select as the initial current best alternative the one that requires the smallest first cost. In most cases the initial current best would be the do nothing alternative. Compare the initial current best alternative and the first challenging alternative. Examine the difference between the two cash flows. If the present worth of the incremental cash flow evaluated at the MARR is greater than zero, the challenger becomes the new current best alternative. If the present worth is less than or equal to zero, the current best alternative remains unchanged and the challenger in the comparison is eliminated from consideration. The new challenger is the next alternative.
• Repeat the comparisons of the challengers to the current best alternative until no alternatives remain.

This method is more time consuming than the present worth on total investment criterion. But both criteria lead to the same solution. In fact,

$$PW(i)_B - PW(i)_A = PW(i)_{B-A}$$

$$AE(i)_B - AE(i)_A = AE(i)_{B-A}$$

$$FW(i)_B - FW(i)_A = FW(i)_{B-A}$$

c. Rate of Return on Incremental Investment

This decision criteria uses the same type of incremental analysis as the prior criterion. The only difference is the decision rule in the third step. For this method, the rule should be

- The increment of investment is considered desirable if the rate of return resulting from the increment is greater than the MARR. that is, $i^*_{B-A} > MARR$.

D.   Applying Decision Criteria When Money Is Limited

Up to now the decision criteria have been applied to sets of mutually exclusive alternatives and it has been assumed that sufficient money is available to undertake all of the proposals. We now assume that the total money available for investment is a fixed amount that restricts the number of feasible alternatives.

No changes at all are required in order to include the budget constraint in the decision process. All that is required is that the proposals be rearranged into mutually exclusive alternatives. Next, those alternatives that have a first cost that exceeds the budget amount must be dropped from consideration. Everything else remains the same.

E.   Comparison of Alternatives with Unequal Service Lives

To calculate the present worth for cash flows of unequal duration is incorrect. When comparing alternatives with unequal lives the principle that all alternatives under consideration must be compared on the same time span is basic to sound decision making. There are two basic approaches that can be used so that alternatives with different lives can be compared over an equal time span.

1. Study Period Approach

This approach confines the consideration of the effects of the alternatives being evaluated to some study period that is usually the life of the shortest-lived alternative. The costs occurring after the study period are disregarded, since the equivalent costs are being compared only for the study period indicated.

2. Estimating Future Alternatives

The second approach is to estimate the future sequence of events that are anticipated for each alternative being considered so that the time span is the same for each alternative. Two methods are used to accomplish this.

a. Expanding Short Life to Long Life

Here we have explicit consideration of future alternatives over the same time span.

b. Common Multiple Life

We assume that an investment opportunity will be replaced by an identical alternative until a common multiple of lives is reached. The annual equivalent comparison should be used since it is computationally the most efficient approach. Since the cash flows for each alternative consist of identical repeated cash flows, it is only necessary to calculate the annual equivalent for the original alternative.

IV. BENEFIT–COST ANALYSIS

With the same data used, the decision reached by comparing benefits with costs is the same decision reached by other methods. But the same proposed project may have several different values of the benefit–cost ratio depending upon whether certain adverse items are subtracted from benefits or added to costs. In general, relevant consequences to the general public should be classified as benefits or disbenefits. Consequences involving disbursements by governmental units should be classified as costs.

In comparing projects, the larger benefit–cost ratio prevails. But you can see that what I call a disbenefit, you could call a cost. Therefore your ratio calculation could differ from mine.

V. DEPRECIATION AND DEPRECIATION ACCOUNTING

A. The Value–Time Function

The pattern of the future value of an asset should be predicted. It is usual to assume that the value of an asset decreases yearly in accordance with one of several mathematical functions. Once a value–time function has been chosen, it is used to represent the value of the asset at any point during its life.

Accountants use the term book value to represent the original value of an asset less its accumulated depreciation at any point in time. The book value at the end of any year is equal to the book value at the beginning of the year less the depreciation expense charged during the year.

In determining book value the following notation is used:

P = first cost of the asset,
F = estimated salvage value,
B_t = book value at the end of year t,
D_t = depreciation charge during year t, and
n = estimated life of the asset.

The following sections are concerned with methods for determining D_t for $t = 1$, $2, \ldots, n$.

1.   Straight-Line Method of Depreciation

This method assumes that the loss in value of the asset is directly proportional to its age. So

$$D_t = \frac{P - F}{n}$$

and

$$B_t = P - t\left(\frac{P - F}{n}\right)$$

2.   Declining-Balance Method of Depreciation

This method assumes that the annual cost of depreciation is a fixed percentage of the book value at the beginning of the year. The ratio of the depreciation in any one year to the book value at the beginning of the year is a constant k. This method gives

$$\text{book value at end of } t \text{ years} = B_t = P(1 - k)^t$$

$$\text{book value at end of } n \text{ years} = B_n = P(1 - k)^n$$

There are two ways of using this method. Knowing the original cost P, the salvage value F, and the useful life n, we use

$$F = B_n = P(1 - k)^n$$

to calculate k and then use

$$B_t = P(1 - k)^t$$

to calculate the book value at the end of t years. We shall call this the general Matheson method.

The usual case, though, is the *double-declining balance*. For this case, the k used is double the depreciation rate of the straight-line method. For the straight-line method the depreciation rate is

$$\frac{D_t}{P - F} = \frac{1}{n}$$

So, for the double-declining-balance method

$$k = \frac{2}{n}$$

and

$$B_t = P\left(1 - \frac{2}{n}\right)^t$$

3. Sum-of-the-Years-Digits Method of Depreciation

For this method, the digits corresponding to the number of each year of life are listed in reverse order, from n to 1. The depreciation factor for any year is the reverse digit for that year divided by the sum of the digits. Since the sum-of-the-years digits is $n(n + 1)/2$, the depreciation factor for any year t is

$$D_t = 2(P - F)\frac{(n - t + 1)}{n(n + 1)}$$

and the book value (after considerable manipulation) becomes

$$B_t = P - \frac{t(2n - t + 1)}{n(n + 1)}(P - F)$$

4. Sinking-Fund Method of Depreciation

This method assumes that a sinking fund is established to accumulate funds for replacement. The total depreciation that has taken place up to any given time is assumed to be the uniform annual sinking-fund deposit plus the interest on the imaginary accumulated fund. This method (unlike the others) preserves the invested capital. So, with this method, the book value at the end of t years will be

$$B_t = P - (P - F) \cdot A/F(i, n) \cdot F/A(i, t)$$

B. Capital Recovery with Return—All Methods of Depreciation

We have seen that the straight-line, declining-balance, sum-of-the-years-digits, and sinking-fund methods of depreciation all lead to different value-time functions for book value. But if the retirement of any asset takes place at the age predicted and the book value equals the estimated salvage value, the depreciation amount and interest on the undepreciated balance can be shown to be equivalent to the capital recovery with a return. Thus, for any method of depreciation, the capital recovery with return is

$$(P - F) \cdot A/P(i, n) + Fi$$

Depreciation is a cost of production. In economic analysis dealing with physical assets, it is necessary to compute the equivalent annual cost of capital recovered plus return, so that alternatives involving competing assets may be compared on an equivalent basis. Regardless of the depreciation model chosen to represent the value of the asset over time, the equivalent annual cost of capital recovered and return will be

$$CR(i) = (P - F) \cdot A/P(i, n) + Fi$$

VI. EXPECTED VALUE DECISION MAKING†

If probability distributions are used to describe the economic elements that make up an investment alternative, the expected value of the cost or profit can provide a reasonable basis for comparing alternatives. The expected profit or cost of a proposal reflects the long-term profit or cost that would be realized if the investment were repeated a large number of times and if its probability distribution remained unchanged. Thus, when large numbers of investments are made it may be reasonable to make decisions based upon the average or long-term effects of each proposal. Of course, it is necessary to recognize the limitations of using the expected value as a basis for comparison on unique or unusual projects when the long-term effects are less meaningful.

To see how these ideas can be useful for decision making, consider the following game. A coin will be tossed twice. If no heads occur, $100 will be lost. If one head occurs, $40 will be won. If two heads occur, $80 will be won. What is the expected value of the random variable, profit from game?

For a discrete random variable x, the expected value is defined as

$$E(x) = \sum_x xP(x)$$

where $P(x)$ is the probability of x occurring. Now if we toss a coin twice, simple reasoning leads to the following probabilities:

$$\text{no heads}\quad P(x) = 1 \text{ out of } 4 = 0.25,$$
$$\text{two heads}\quad P(x) = 1 \text{ out of } 4 = 0.25, \text{ and}$$
$$\text{one head}\quad P(x) = 2 \text{ out of } 4 = 0.50.$$

Therefore

$$E(G) = -\$100(0.25) + \$40(0.50) + \$80(0.25) = \$15$$

If one were able to participate in a large number of such games, the expected or average winnings per bet over the long run would be approximately $15.

Because most industries and governments are generally long lived, the expected value as a basis for comparison seems to be a sensible method for evaluating investment alternatives under risk. The long-term objectives of such organizations may include the maximation of expected profits or minimization of expected costs. If it is desired to include the effect of the time value of money where risk is involved, all that is required is to state expected profits or costs as expected present worth, expected annual equivalents, or expected future worths.

---

† This section is based upon H. G. Thuesen, W. J. Fabrycky, and G. J. Thuesen, *Engineering Economy*, 5th ed., Prentice-Hall, Englewood-Cliffs, New Jersey, 1977, pp. 430–432. Used by permission of Prentice-Hall.

As an example of decision making using expected values, suppose that a community has a water treatment facility located on the flood plain of a river. The construction of a levee to protect the facility during periods of flooding is under consideration. Data on the costs of construction and expected flood damages are shown in Table E1.

TABLE E1. Probability and Cost Information for Determining Optimum Levee Size

A	B	C	D	E
x(feet)	Number of Years River Maximum Level Was x ft Above Normal	Probability of River Being x ft Above Normal	Loss If River Level Is x ft Above Levee ($)	Initial Cost Building Levee x ft High ($)
0	24	0.48	0	0
5	12	0.24	100,000	100,000
10	8	0.16	150,000	210,000
15	3	0.06	200,000	330,000
20	2	0.04	300,000	450,000
25	1	0.02	400,000	550,000
	50	1.00		

Using historical records that describe the maximum height reached by the river during each of the last 50 years, the frequencies shown in column B were ascertained. From these frequencies are calculated the probabilities that the river will reach a particular level in any one year. The probability for each height is determined by dividing the number of years for which each particular height was the maximum by 50, the total number of years.

The damages that are expected if the river exceeds the height of the levee are related to the amount by which the river height exceeds the levee. These costs are shown in column D. It is observed that they increase in relation to the amount the flood crest exceeds the levee height. If the flood crest is 15 ft and the levee is 10 ft the anticipated damages will be $100,000, whereas a flood crest of 20 ft for a levee 10-ft high would create damages of $150,000.

The estimated costs of constructing levees of various heights are shown in column E. The company considers 12% to be their minimum attractive rate of return and it is felt that after 15 yr the treatment plant will be relocated away from the floodplain. The company wants to select the alternative that minimizes its total expected costs. Since the probabilities are defined as the likelihood of a particular flood level in any one year, the expected equivalent annual cost is an appropriate choice for the basis for comparison.

An example of the calculations required for each levee height is demonstrated for two of the alternatives.

Five-foot levee:

$$A/P(12, 15)$$

annual investment cost = $100,000(0.1468)   = $14,682

expected annual damage = (0.16)($100,000)
$$+ (0.06)($150,000)$$
$$+ (0.04)($200,000)$$
$$+ (0.02)($300,000) = \underline{39,000}$$

total expected annual cost $53,682

Ten-foot levee:

$$A/P(12, 15)$$

annual investment cost = $210,000(0.1468)   = $30,828

expected annual damage = (0.06)($100,000)
$$+ (0.04)($150,000)$$
$$+ (0.02)($200,000) = \underline{16,000}$$

total expected annual cost $46,828

The costs associated with the alternative levee heights are summarized in Table E2. The levee height that minimizes the total expected annual cost is the levee that is 10 ft in height. The selection of a smaller levee would not provide enough protection to offset the reduced construction costs, while a levee higher than 10 ft requires more investment without proportionate savings from expected flood damage. The use of expected value in determining the cost of flood damage is reasonable in this case, since the 15-yr period under consideration allows time for long-term effects to appear.

TABLE E2. Summary of Annual Construction and Flood Damage Costs

Levee Height (ft)	Annual Investment Cost ($)	Expected Annual Damage ($)	Total Expected Annual Costs ($)
0	0	80,000	80,000
5	14,682	39,000	53,682
10	30,828	16,000	46,828
15	48,450	7,000	55,450
20	66,069	2,000	68,069
25	80,751	0	80,751

PROBLEMS

A. A manufacturer has purchased a machine for $80,000. Its useful life is 5 yr, at which time it can be sold for $5000. The hourly costs are $7.50/hr; the annual costs are $1000/yr. This machine is used to make castings. Each casting requires about 100 min to produce. The material for each casting is $8.00. If the cost of money is 8%, calculate the cost per unit to make each casting if (a) 120, (b) 220, and (c) 550 castings are made each year.

B. We have at least a million dollars for investments if we can receive a return of 15%/yr. Three different proposals for modifying our assembly line have been made. Which proposal should be accepted?

Proposal	A	B	C
Investment ($)	140,000	165,000	190,000
Annual savings ($)	21,750	26,750	29,750

C. We have two alternatives for our proposed new grease canning line. Plan A uses an electrically powered belt that costs $1500 with a 5-yr salvage value of $300. The power cost is $1.00/hr of use and maintenance should run $150/yr.

Plan B uses a reconditioned gas engine costing $500 with no salvage value. Gas and oil costs are estimated at $0.60/hr and the operator (usually the janitor) is paid $3/hr. Because of its age, we assume maintenance costs to be $0.25/operating hr.

We assume that the canning line will be needed for 5 yr. Management requires an overall rate of return of 12% before taxes.

Determine the hours of line use per year for the two plans to break-even. Solve by equivalent annual cost methods before taxes.

D. After our refinery was built on an island in the river, we were told that the island was occasionally under water. Corps of Engineers data indicate that the chances of a flood are about one in seven each year. The likely flood damage would cost $25,000.

A levee could be built at a cost of $60,000 with a useful life of 28 yr with no salvage. If money is available at 10%, should we build the levee?

E. The chairman will be visiting the plant. We must paint. Paint A costs $7.00/gal and covers 350 sq ft/gal. The manufacturer claims that it will last 4 yr and can be applied at a rate of 100 sq ft/hr.

Paint B costs $10/gal and covers 500 sq ft/gal. It will last for 5 yr and can be applied at a rate of 125 sq ft/hr. If we pay the painter $4/hr, which paint should we use. Money is available at 10%.

F. We wish to invest some of our excess profits in a new branch office. We have money enough for one. It would cost $150,000 to set up a Cleveland office. Annual revenues would be $68,000 and annual costs $58,000. It would cost $180,000 to open a Pittsburgh office. Annual costs would be $50,000 and annual revenues $75,000. If we assume no salvage value after 20 yr for either office, which branch would be the most profitable (i.e., yields greatest return on investment)?

G. We have the choice of building a new plant in one or two stages. We have cost estimates for both choices, based upon 1980 costs. Which method, based upon the following data, should be recommended?

We can build the complete plant or we can build it in two stages. The increased capacity is not needed until 20 yr from now. We can borrow money at 10%. Construction costs will rise at 6% per year. Maintenance costs after 20 yr are assumed equal for the two projects.

	Capacity (lb/yr)	Annual Maintenance ($)	Construction Costs (1980) ($)
First stage	100,000	80,000	14,000,000
Second stage	175,000		12,000,000
All at once	175,000	85,000	21,000,000

H. We wish to build an R & D office with a first cost of $400,000, annual operating and maintenance costs of $40,000/yr, and a salvage value of $60,000 at the end of 50 yr. Assume an interest rate of 10%.

(a) What is the present value of this investment if the planning horizon is 50 yr?

(b) If the building replacement will have the same first cost, life, salvage value, and operating costs as the original, what is the capitalized cost of perpetual service?

I. The gas company data shows the following annual frequency of leaks and cost characteristics for their processing plant:

Number of leaks	0	1	2	3	4
Probability of this number during year	0.3	0.2	0.2	0.2	0.1
Cost of this number of leaks to gas company	$0	$1500	$2500	$5000	$11,000

With the installation of new piping, the above probabilities would be changed to

$$P(0) = 0.75, \qquad P(1) = 0.15, \qquad P(2) = 0.10, \qquad P(3) = P(4) = 0$$

If the new piping will have a life of 20 yr, no salvage, and an annual maintenance of $500, determine the amount that the gas company could afford to spend on the project. Assume a 12% interest rate.

PROBLEM-SOLVING STRATEGIES

A. Find the yearly cost of the equipment and then add to it the annual costs. These costs will be the same regardless of the number of castings produced. Then for each different number of castings, determine the hours required to produce them. Calculate the total hourly costs for each number of castings. The total annual cost to produce a number of castings is this cost added to the other yearly costs. Dividing by the number of units provides the answer. This is a simple, straightforward problem.

B. This is a mutually exclusive alternatives problem. The straightforward method is to list investments by ascending first costs, comparing percentage return on both first cost and on incremental investments.

C. For this problem, add up the total annual costs for both alternatives, letting H be the hours of line use per year. The break-even point occurs when the two annual costs are equal.

D. The secret to the solution of this problem is to decide how to handle the probability of a flood. A simple way to do it is to assume that the yearly cost of a flood that occurs once every 7 yr is $\frac{1}{7}$ of the total cost of a flood. This value is then compared to the annual cost of building a levee. This concept fits with the life of the levee being 28 yr — a constant multiple of seven.

E. This is a break-even problem. We must determine the annual cost for each paint. Data looks as if the initial cost will be in dollars per square ft year. Note that one paint will last 4 yr, the other 5 yr.

F. We can determine the annual profit for each office. Since we know the investment needed to set up each office, we can calculate a rate of return for each, based on

$$A/P(\%, 20) = \text{profit}/\text{investment}$$

The larger rate of return is chosen.

G. Although there are several approaches to this problem, calculating present worth is probably the best. We must remember to update the construction cost of the second stage by 6%/yr. Use the plan that has the lower present worth.

H. From the data given, calculate the present worth, assuming a life of 50 yr. Use the relationship between present worth and capitalized annual cost to get the capitalized cost from the present worth.

I. We must know how to handle probabilities in economic analysis to do this problem. In this case, it is simple. The total cost is just the sum of all individual costs times occurrence probabilities. With this in hand, we must calculate the annual cost of new piping, assuming C is the initial cost. A break-even analysis completes the problem.

SOLUTIONS

A. (a) Cost to make 120 castings

time = $(120 \times 100$ min$)/60$ min/hr = 200 hr

yearly cost of equipment = ($80,000 - 5000)A/P(8\%, 5)$ = $18,784.50
$$[A/P = 0.25046]$$

fixed cost = $5000 \times i[i = 0.08]$ =   400.00

annual costs   =   1,000.00

hourly costs = $7.50/hr $\times$ 200 hr   =   1,500.00

materials = 8.00×120 =   960.00

total cost   = $22,644.50

cost/unit = $22,644.50/120   = $188.70
                                                                    each

(b) Cost to make 220 castings

time = $(220 \times 100)/60 = 366.67$ hr

yearly cost   = $18,784.50

fixed cost   =   400.00

annual costs   =   1,000.00

hourly costs = $7.50/hr $\times$ 366.67 hr   =   2,750.03

materials = 8.00×220 =   1,760.00

total cost   = $24,694.53

cost/unit = $24,694.53/220   = $ 112.25

(c) Cost to make 550 castings

$$time = (550 \times 100)/60 = 916.67 \text{ hr}$$

yearly cost	= $18,784.50
fixed cost	= 400.00
annual cost	= 1,000.00
hourly costs = $7.50/hr × 916.67 hr	= 6,875.03
materials = $8.00 × 500	= 4,400.00
total cost	= $31,459.53
cost/unit = $31,459.53/550	= $ 57.20

B. Need a return of at least 15%

Proposal	Invest-ment ($)	Savings ($)	% Return	Δ Invest-ment ($)	Δ Savings ($)	Δ% Return
A	140,000	21,750	15.5	—	—	—
B	165,000	26,750	16.2	25,000	5000	20.0
C	190,000	29,750	15.7	25,000	3000	12.0

Proposal A wins against a simple 15% investment; B wins against A because both $\%$ return and $\Delta\%$ return over A are above 15%, and C loses against B because the incremental return is less than 15%.
   So we accept proposal B.

C. Plan A—annual cost

capital recovery = ($1500 − 300)$A/P$(12\%, 5)$		
$[A/P = 0.27741]$	=	$332.89
interest on salvage = $300i[i = 0.12]$	=	36.00
power cost = $1.00H$	=	H
maintenance = $150/yr	=	150.00
total cost	=	$518.89 + H$

Plan B—annual cost

capital recovery = $500 $A/P(12\%, 5)[A/P = 0.27741] = \$138.71$
gas and oil = $0.60 H$ = $0.60 H$
operator = $3.00 H$ = $3.00 H$
maintenance = $0.25 H$ = $0.25 H$
     total cost   = $\$138.71 + 3.85 H$

break-even = $\$518.89 + H = \$138.71 + 3.85 H$

$380.18 = 2.85 H$; $H = 133.4$ hr

Above 133.4 hr favors plan A.

D. Likely cost of a flood = $\frac{1}{7}$ ($25,000) = $3571.43 each year

Annual cost of building levee = $60,000 $A/P(10\%, 28)$
                              = $60,000 × 0.10745 = $6447

So we do *not* build the levee.

E. Annual cost—paint A $[A/P(10\%, 4) = 0.31547]$

$$\left[\frac{\$7.00}{gal} \frac{1\ gal}{350\ ft^2} + \frac{\$4}{hr} \frac{1\ hr}{100\ ft^2} \right] \times [0.31547]$$

$$= \$0.060/ft^2 \times 0.31547 = \$0.01893/ft^2\ yr$$

Annual cost—paint B $[A/P(10\%, 5) = 0.26380]$

$$\left(\frac{\$10}{gal} \frac{1\ gal}{500\ ft^2} + \frac{\$4}{hr} \frac{1\ hr}{125\ ft^2} \right) \times [0.26380]$$

$$= \$0.052/ft^2 \times 0.26380 = \$0.01372/ft^2\ yr$$

So we use paint B.

F. Cleveland:

annual revenue less annual cost = $68,000 − $58,000 = $10,000
capital recovery = $150,000 A/P $(i, 20)$
$A/P(i, 20) = \$10,000/\$150,000 = 0.06667$

$A/P(2\frac{1}{2}\%, 20) = 0.06415$; $A/P(3\%, 20) = 0.06722$

Pittsburgh:

annual revenue less annual cost = $75,000 − $50,000 = $25,000
capital recovery = $180,000 $A/P(i, 20)$
$A/P(i, 20) = \$25,000/\$180,000 = 0.13889$

$$A/P(12\%, 20) = 0.13389; \qquad A/P(13\%, 20) = 0.14235$$

then

$$i \cong 12\% + \frac{0.13889 - 0.13388}{0.14235 - 0.13388}$$

$$\cong 12\% + 0.59 \qquad \text{or} \qquad i = 12.59\%$$

So we open a branch in Pittsburgh.

G. First plan—two stages   1st today, 2nd 20 yr hence
present worth (1st stage)$[P/A(10\%, 20) = 8.514]$

$$\$14,000,000 + \$80,000P/A(10\%, 20) = \$14,681,120$$

present worth (2nd stage)$[F/P(6\%, 20) = 3.2071; P/F(10\%, 20)$
$= 0.1486]$

$$\$12,000,000 \; F/P(6\%, 20) \; P/F(10\%, 20) = \$5,718,901$$

total present worth—first plan = $20,400,021
second plan—build all today
present worth $[P/A(10\%, 20) = 8.514]$

$$\$21,000,000 + \$85,000 \; P/A(10\%, 20) = \$21,723,690$$

Thus we should use first plan with two stages.

H. (a)   Present worth $[P/A(10\%, 50) = 9.915; P/F(10\%, 50) = 0.0085]$

$\$400,000 + \$40,000 \; P/A(10\%, 50) - \$60,000 \; P/F(10\%, 50) = \$796,090$

(b)   capitalized annual cost $[P/A(10\%, 50) = 9.915]$

$$\frac{\text{PW}}{i \cdot P/A(10\%, 50)} = \frac{\$796,090}{0.10 \times 9.915} = \$802,915$$

I. Expected leaks cost of old piping

$$= \sum_{p} \text{cost } (p) \cdot P(p)$$

$$= 0.3(\$0) + 0.2(\$1500) + 0.2(\$2500) + 0.2(\$5000) + 0.1(\$11,000)$$

$$= \$2900$$

expected leaks cost of new piping
$$= 0.75(\$0) + 0.15(\$1500) + 0.10(\$2500)$$
$$= \$475$$

Equate annual equivalent cost of old system to new system $[A/P(12\%, 20) = 0.13388]$

$$\$2900 = \$475 + \$500 + \text{cost} \cdot [A/P(12\%, 20) = 0.13388]$$
$$= \$975 + 0.13388 \text{ cost}$$

or

$$\text{cost} = (\$2900 - \$975)/0.13388 = \$14{,}378.55$$

We could profitably afford to pay up to $14,378.55 for the new system.

INDEX